QUANTUM PHYSICS FOR BEGINNERS

THE EASY GUIDE TO LEARN QUANTUM PHYSICS AND
THE THEORY OF RELATIVITY

By
Cary Hanson

© Copyright 2020 by Cary Hanson
All rights reserved.

This document is geared towards providing exact and reliable information with regards to the topic and issue covered. The publication is sold with the idea that the publisher is not required to render accounting, officially permitted, or otherwise, qualified services. If advice is necessary, legal or professional, a practiced individual in the profession should be ordered.

- From a Declaration of Principles which was accepted and approved equally by a Committee of the American Bar Association and a Committee of Publishers and Associations.

In no way is it legal to reproduce, duplicate, or transmit any part of this document in either electronic means or in printed format. Recording of this publication is strictly prohibited and any storage of this document is not allowed unless with written permission from the publisher. All rights reserved.

The information provided herein is stated to be truthful and consistent, in that any liability, in terms of inattention or otherwise, by any usage or abuse of any policies, processes, or directions contained within is the solitary and utter responsibility of the recipient reader. Under no circumstances will any legal responsibility or blame be held against the publisher for any reparation, damages, or monetary loss due to the information herein, either directly or indirectly.

Respective authors own all copyrights not held by the publisher.

The information herein is offered for informational purposes solely, and is universal as so. The presentation of the information is without contract or any type of guarantee assurance.

The trademarks that are used are without any consent, and the publication of the trademark is without permission or backing by the trademark owner. All trademarks and brands within this book are for clarifying purposes only and are the owned by the owners themselves, not affiliated with this document

TABLE OF CONTENTS

INTRODUCTION — 5

WHAT IS QUANTUM PHYSICS — 9

QUANTUM PHYSICS - THE LOCALIZATION OF MANIFESTATION! — 13

QUANTUM THEORY - AN OVERVIEW OF THE MYSTIFYING SCIENCE — 19

QUANTUM PHYSICS AND LAW OF ATTRACTION — 21

QUANTUM THEORY — 23

QUANTUM PHYSICS FOR BETTER HEALTH — 27

QUANTUM PHYSICS AND YOU — 30

QUANTUM PHYSICS - THE DISCOVERY THAT SCIENTIFICALLY DEMOLISHED MATERIALISM — 34

THE QUANTUM DIMENSION — 41

THE RELATION BETWEEN WAVES AND PARTICLES — 43

WAVE-PARTICLE DUALITY — 49

THE BUILDING BLOCKS OF MATTER AND WAVE-PARTICLE DUALITY — 55

WHY MAX PLANCK IS CALLED THE FATHER OF QUANTUM PHYSICS — 58

- LAWS OF QUANTUM PHYSICS — 61
- QUANTUM FIELD THEORY AND STANDARD MODEL — 67
- QUANTUM FIELD THEORY — 70
- EINSTEIN'S THEORY OF RELATIVITY — 88
- WHAT IS THE THEORY OF RELATIVITY? — 91
- SIX THINGS EVERYONE SHOULD KNOW ABOUT QUANTUM PHYSICS — 95
- RELATIVITY, QUANTUM PHYSICS, AND STRING THEORY — 102
- BASICS ON ANGULAR MOMENTUM ON A QUANTUM LEVEL — 105
- THE VERY ESSENCE OF SPACE-TIME AND GRAVITY — 110
- CONCLUSION — 112

INTRODUCTION

Quantum mechanics, commonly called quantum physics, is the relationship between energy and matter. The word 'quantum' is Latin for 'how much.' Mechanics refers to a unit in which quantum theory assigns to specific physical quantities in minimal quantities as a measure. In essence, quantum expressions are generally visualized and studied subatomically with subatomic particles.

Subatomic particles are small. If an atom was the size of a house, the subatomic molecule would be the size of a drop of chewing gum in the kitchen cabinet of that house.

There were a few things that needed to occur before the investigation of quantum mechanics flourished. In 1838 was the disclosure of cathode beams and Gustav Kirchoff published a statement in 1850 on the problem of 'blackbody radiation.' So in 1877, Ludwig Boltzmann proposed that the energetic states of a physical system could be unconnected.

In 1900 Max Planck developed the theory that energy is radiated and absorbed. He made an equation known as 'Planck's activity consistent.'

Planck is known as the grandad of quantum material science. After his theory was circulated, other scientists took note of it and discovered other theoretical structures until, eventually, quantum mechanics was theorized and studied around the world.

Because of quantum physics, we discovered gravity, we have superconductors and magnetic resonance imaging equipment in hospitals, and now we can even see that time travel is possible.

It all sounds so fantastic, but scientists in the field of quantum mechanics will tell you. It is hard for a most of us to comprehend the connection between subatomic particles and the law of attraction.

During the investigation of quantum mechanics, it was discovered that subatomic particles determine the direction that the earth is turning. Another force moves these particles of physical matter out into the universe.

After some double-blind slit tests using subatomic particles as subjects, it was discovered that they could switch between wave-shaped particles and then back to block shaped particles again. These particles could leave our dimension and enter it again. We also found that these subatomic particles changed deliberately from wave-shaped particles depending on the purpose. So, we found out that when we were testing the particles, we couldn't remove ourselves from the equation. We influenced the particles by thinking about the result.

This is where it becomes confusing and the concept confused Einstein until his death. Understanding particle and wave duality is not easy for most of us to comprehend.

However, one of the theories that emerged from the foundations of quantum physics is that we manipulate the fabric of life by thinking about it. Our thoughts have an expression that comes out and therefore brings us to what we focus on so that it's a reality. This is the law of attraction.

Once you understand what your world is, only then do you begin to understand its true behavior and nature. You, at that point, can change your perspective on it and with your changed observation, you change your formation and, therefore, your physical reality. This is the first step to prosperity.

Every single quantum physicist will agree on one thing ...

The subatomic particles, the energy packets or quantum, are not particles at a certain point in space and time like a table or a chair, but they are just a probability that they could exist at different points in space and time.

The act of our observation converts it into a 'physical' particle at a certain point in space and time. Once we withdraw that attention, it becomes a probability again. Imagine that the sofa in your living room is a sizeable subatomic particle.

This is how it would behave: If you are not in the house and do not think of your sofa, it would 'disappear' and become a probability that it could reappear anywhere in your living room or anywhere else in the universe!

If you come home thinking about sitting on the sofa in a specific place in your living room and looking for the sofa where you want to relax, it reappears!

This seems like a kind of fantasy, but, it's a scientific fact that subatomic particles behave in the same way.

The presence of your sofa is only the result of you seeing it, expecting it and deciding that it is there. It is not a completely independent existence. No matter is an entirely separate existence, regardless of the observer.

The astonishing thing is that all matter consists only of large amounts of these particles. Therefore, all matter behaves precisely as a large group of subatomic particles would.

Quantum physics confirms that a thing can only exist if it is observed. The 'quanta' is organized according to the influence of the mind principle of the observers.

When something is observed, quanta merge into subatomic particles and then into atoms, followed by molecules, until finally, something in the physical world manifests itself as a localized temporal space-time experience that can be perceived through the mediation of our five physical senses, then lead to something that appears to be reliable and is part of what people usually understand as physical reality.

An experiment by modern quantum physicists shows and proves that the entire universe exists through experience. The most fantastic research in quantum physics in recent scientific discovery is probably the double-slit experiment.

Every single thought, as energy, directly and instantly influences the quantum field, whereby 'Quanta' merges into a localized, observable experience event, object, or other influence. This process is the basis for how everyone creates their reality.

Those who understand and comply with universal laws are conscious creators, while others create their life experience by default. As a result, they attribute everything that has been experienced as being a consequence of their unconscious thinking to superstitious beliefs such as luck, fate, chance and fortune. We know that conscious creation is also the basis of the law of attraction and the law of cause and effect!

WHAT IS QUANTUM PHYSICS

What is quantum physics? Put simply, it is physics that explains how everything works: the best portrayal of the kind of particles that make up the issue and the powers with which they communicate.

Quantum physics is the method in which molecules work and how science and scientists work out how they work. We as a whole, move to the quantum tune and all work in the same way. If you need to clarify how electrons travel through a PC chip, how light photons in a sunlight based board are changed over into power or intensified into a laser, or how the sun keeps on consuming, you need to use quantum physics.

The trouble, and for physicists, the enjoyment, begins here. Most importantly, there isn't a single quantum theory. There's quantum mechanics, the hidden scientific system behind it, which was first evolved during the 1920s by Niels Bohr, Werner Heisenberg, Erwin Schrödinger and others. It shows basic things like how the position or energy of a solitary molecule or gathering of fewer particles changes over time.

To see how things work in reality, quantum mechanics should be combined with different components of material science - particularly Albert Einstein's specific theory of relativity, which clarifies what happens when things move rapidly - to what makes quantum field speculations,

Three unique methodologies about quantum fields manage three of the four crucial powers with which matter cooperates: electromagnetism, which clarifies how iotas hold together; the reliable nuclear power that defines the stability of the nucleus in the heart of the atom; and the weak atomic power that explains why some atoms decay radioactively.

Over the past five decades, these three theories have been merged into a dilapidated coalition known as the 'standard model' of particle physics. Despite the impression that this model is held together with tape, it is the most thoroughly tested picture of the essential work of matter that has ever been developed. Its heyday was in 2012 with the discovery of the Higgs boson, the particle that gives all other fundamental particles their mass, the existence of which was predicted in 1964 based on quantum field theories.

Conventional quantum field theories well describe the results of experiments with high-energy particle destroyers, such as CERN's Large Hadron Collider, in which the Higgs was discovered, examine the matter on a microscopic scale. However, if you want to understand how things work in less stressful situations, how electrons move for example, or don't travel through a strong material, making a material a metal, a protector, or a semiconductor, for instance, things get much more confused.

The billions and billions of interactions in these crowded environments require the development of 'effective field theories' that cover some of the essential details. The difficulty in constructing these theories is why many fundamental questions in solid-state physics remain unsolved, for example, because some materials are superconductors at low temperatures they allow currents without electrical resistance, and because we cannot run this experiment at room temperature, we're unable to resolve it.

However, all these practical problems hide a huge quantum secret. Quantum physics predicts surprising things about how matter functions that are inconsistent with how things appear to function in reality. Quantum particles can act like particles that stay still. Or they can act as waves that are distributed in space or at several points at the same time. The way they appear seems to depend on how we measure them. Before we measure them, they seem to have no defined properties at all, which leads us to a fundamental mystery about the nature of underlying reality.

This confusion causes apparent paradoxes like Schrödinger's cat, in which, thanks to an uncertain quantum process, a cat remains dead and alive at the same time. But that's not all; even quantum particles seem to be able to influence each other immediately, even if they are far apart. This truly passionate phenomenon is known as entanglement or, as Einstein (a great critic of quantum theory) has said, 'distant spectral action.' Such quantum forces are entirely foreign to us, but they form the basis for new technologies such as high-security quantum cryptography and high-performance quantum computing.

But nobody knows what it all means. Some people think we have to accept that quantum physics explains the material world with terms that we cannot possibly reconcile with our experience in the larger 'classic' world. Others think there has to be a better, more intuitive theory that we haven't discovered yet.

In all of this, there are several elephants in the room. First of all, there is a fourth fundamental force of nature that quantum theory has so far, not been able to explain. Gravity remains the territory of Einstein's general theory of relativity, a fixed non-quantum theory in which not even particles are involved. Decades of intensive research to put gravity under the umbrella of the quantum and thus to explain all of the basic physics within a 'theory of everything' have failed.

Meanwhile, cosmological measurements show that over 95 percent of the universe consists of dark matter and dark energy, for which we currently have no explanation in the standard model, and dilemmas such as the expansion of the role of quantum physics in the disordered mechanisms of life remain unexplainable. The world is at a certain quantum level, but if quantum physics is the last word in the world, it remains an open question.

QUANTUM PHYSICS - THE LOCALIZATION OF MANIFESTATION!

Quantum physicists speak of electrons or events that are potential rather than real physical units. So there are several potentials until someone looks, and then forces the universe to make a decision, so to speak, a determination as to which potential to locate and update. All existence is an unlimited quantum field of energy, a sea of infinite possibilities that are all waiting to happen!

The mind creates and controls reality. Here's how the law of attraction works. We get what we focus on most of the time. The viewer creates reality by merely observing.

The mind, regardless of the structure, contains images. And every image that is held firmly in someone's brain in any form, must come out.

Whenever the mind forms a mental image or an image of anything, it becomes 'one' with the infinite universal consciousness, and the developed image is then outsourced to the physical world as a single space-time event. For an image to manifest itself, however, there must be no other contradictory thoughts to abolish the power of manifestation of the image contained in mind.

Another property of the quanta is that they are multidimensional. You are now scientifically seeing that our universe is multifaceted, even though our senses understand length, width, height and time as the only dimensions. However, our souls are multidimensional. Listen to your soul and your feelings.

The physical world is comprised of thoughts and vitality. Many quantum physicists, including Einstein, have shown that all physical matter is made up of energy packets that are not related to space and time.

This energy field has no precisely defined limits. The universe is vast, timeless and limitless. Science has also shown that the mind has no limits. All thoughts are 'connected' to a field of spiritual energy. You are taller and much more powerful than you think.

Whatever you want, you already have everything. It has been said that it will be given to you before asking. Science is slowly dealing with quantum physics to prove that this is scientifically true. The infinite intelligence of formless substance, the possibility at the quantum level, and our ability to influence this field give us the experience of 'having everything.'

You already have all the riches beyond your wildest dreams. We are now starting to find out on a larger scale, both scientifically and spiritually. You have it. Maybe you're not experiencing it right now, but you could if you believe that you do.

Having and experiencing are two different things. An easy way to explain it is that you can climb Everest or paraglide, but you may not have experienced alls aspects of your abilities. All you have to do is try and you'll be able to do it.

The quantum field can create an infinite number of structures, forms and experiences and has already done so. The pages in this book are just one of the things created. The words that you're reading are also structures already created. The next thought you have will also be one of these things.

But did you ever expect that you would experience these words on these pages? Your desire to find such words made them appear in your hands. They have always existed. But out of the passion that you and many others like you sent to the universe, I was inspired to give you these answers!

It is not necessary to predict exactly how things will develop. All you have to do is desire, understand and know that it is possible and will be arranged to come to you.

In our life, we simply move our consciousness to experience aspects of ourselves that we have always had, in a universe that has everything we could wish for. Even what we do not imagine exists.

Many physicists working on subatomic particles discover some interesting things about our universe, for example, they found that two particles separated by space and time can be 'invisibly connected' and function synchronously. They also found that the world we live in seems built to get to know each other.

This seems to have been done by 'cutting the whole' into at least two halves, with one half conditioned to see and the other also to see. Who is conditioned to understand is, therefore, under the illusion of separation, from what is conditioned to see? It is a necessary illusion. But, in reality everything is 'one'.

Sir Isaac Newton thought of nature as a machine that describes the pricing laws that govern the operation of this machine. Science has seen the universe as a three-dimensional space through which physical objects move according to immutable laws. We learn from these laws that brought us modern technology from steam engines to spacecraft.

We had to dig deep into the heart of the atom when it was discovered in 1890. The world inside the nucleus, which was coined in ancient Greece, means 'indivisible unity,' helped us to discover radioactivity, and it showed how the atom was divisible. Everything is energy. We all know this term: $E = mc^2$. This energy corresponds to the mass for the speed of light squared.

We have found that energy and matter are connected, which can be converted back and forth. We use a study called quantum physics as a study of how the world works on the smallest scales at a much lower level than the atom. There are various elusive small energy packets that physicists call quantum.

Everything is energy, be it a rock, a planet or a glass of water, and everything that can be touched and has a taste or smell. All things are made up of molecules, made up of atoms combined with protons, electrons and neutrons that create this vibrating energy package.

Sir Isaac Newton thought of nature as a machine that describes the pricing laws that govern the operation of this machine. Science has seen the universe as a three-dimensional space through which physical objects move according to immutable laws. We learn from these laws that brought us modern technology from steam engines to spacecraft.

When you build a vision card and insert images of what you want on it, the boundaries between the physical world and the world of our thoughts also disappear. Your thoughts travel in the universe because there is no absolute difference between matter and energy. These small energy packages have the unrecognizable ability to influence each other with a property called entanglement.

We came across the unified field theory or a 'theory of everything' when you know you know the spirit of God. The fact is that thought influences matter, which becomes a small package of possibilities. When we try to record information about the position of a proton, it is not possible to determine its speed or trajectory. If you find out the rate, you can't determine the exact location. We now know that physical matter is not yet. With this new understanding, the reality is not a solid substance, but a field of potential. If we have ideas or sketches of an object that are still waiting to be materialized when we finish the product in production and measure its properties, it would soon dematerialize when it is measured or observed. This act influences the behavior of the particles.

The laws of attraction

The quantum universe has no accidents, no coincidences; Every particle and every action is taken into consideration. Laws are demanding. The laws that regulate the movements of subatomic particles and solar systems also restrict our thoughts and feelings, our families and our careers.

Quantum physics

Two worlds: The visa and the invisible. The key comes from the invisible world because the unseen world is more powerful than the seen world. It is the large and hidden part of the iceberg that most of us do not know 99 percent of the time, while the material world that we consider 'real' is only the small tip that is just above the surface of our consciousness. Pushes are what we don't see.

It all starts as an idea. Here's how the universe works with your design. Everything in the physical world is made up of atoms. Atoms are made of energy. Our consciousness produces power. We create from the non-physical level and transform what we cannot see into what we can.

Newton utilized the standard of circumstances and logical results, how it works in the physical and mechanical world, as the reason for his law of thermodynamics. Newton did not understand that matter, energy and consciousness are not separate areas, but all merely different frequencies along the same continuum.

We currently realize that the circumstances and logical results guideline applies not exclusively to the mechanics of issue, yet to the mechanics of everything, including our contemplations. Every thought you make emits a specific frequency, and that frequency triggers a response from the quantum universe, surely like a vibrating hammer on the surface suggests. In Eastern Karma, in the west, the golden rule is; the essence of the principle is that you have a cause in your life and business. With the opposite side of any situation, you have more opportunities to create events and circumstances in your life than you ever thought possible.

The resonance principle means that energy in a specific model or frequency resonates with any other form of energy in a similar model. The law of attraction is simply the effect of resonance together with cause and effect. Thoughts create events and circumstances that have the same pattern and therefore resonate with those thoughts.

To create your successful situation, you have to start with the part that no one else sees, the hidden part of the iceberg. The seed of your company is a vision. There is a gestation period. It has a law. Here is the law of pregnancy. It is explained that there is a gestation period or a fixed incubation period for each seed, a certain period taken by a particular seed to settle, before it can develop from the project to a fully realized physical form. There is a zero-point field, the quantum vacuum, which knows everything, all-powerful and is all capable.

QUANTUM THEORY - AN OVERVIEW OF THE MYSTIFYING SCIENCE

Quantum theory is the most important, exciting, challenging, and even mystifying scientific domain and is much more than outlandish. Today it is also the most impressive theory in the world. The conjecture tells us that we may be profoundly mistaken in our considerations of what reality is.

The theory was initially called quantum mechanics because it was believed that there should be some universal laws relating to the activity of atomic particles and energy quanta that resemble the mechanics of macroscopic objects such as that of large planets. The theory tries to represent the behavior of microscopic entities, generally the size of atoms, in the same way as Einstein's theory of relativity, to illustrate the laws of larger, everyday entities. It is used in many activities, including television and personal computers, and even describes the nuclear practices that occur in and around the stars.

Quantists let us live in many dimensions that are arranged in the middle of 'waves of probability' and hidden 'virtual particles' that move in and out of creation. They also verbally express that one day, we could slip through wormholes in the universe to look around in other cosmoses or travel back in time. In simple terms, however, quantum theory is the analysis of jumps from one energy level to another, since it refers to the structure and behavior of atoms.

In 1905 Albert Einstein suggested that light is a particle and not a wave, which questioned 100 years of research. He suspected that not only the energy but also the radiation itself was quantized identically.

This is the source of Einstein's well-known challenge that 'God does not roll the dice.' Einstein certainly could not consider it as a closed science, since quantum mechanics generally 'only' provides probabilities on how unique particles would react and cannot process specific certainties. For this reason, despite his many new approaches, Einstein has never been able to abandon the purpose of pre-quantum science in order to predict the cosmos like a clock. Quantum science is not an unfinished science, as Einstein had imagined, but a very progressive science, since it recognizes that science can at best assume expectations from the reaction of certain divisions into complicated techniques.

Without a doubt, Albert Einstein's quantum theory and relativity theory form the basis for today's physics, with almost all individuals believing it to be essentially the theory of the invisible sphere, tiny particles and huge accelerators. For most people, however, it is a slogan for puzzling and unfathomable science. However, it has a much wider field than just narrowing the ball and may be suitable for techniques where many unique sections work together and influence each other at the same time.

QUANTUM PHYSICS AND LAW OF ATTRACTION

If you've heard of the law of attraction, you've probably come across mentions of quantum physics. At first glance, these two arguments seem to separate the worlds. But if so, why do people mention quantum physics and the law of attraction together as if there was a deep connection between them? Let's take a look at why these two seemingly different topics are considered and mentioned at the same time.

What Exactly Is Quantum Physics?

Quantum physics is the branch of physics that specifically deals with quanta. The quanta are nothing but tiny or indivisible parts of energy. Quantum physics is based on some crucial, fundamental facts. First, it makes clear that the quantum world is completely different from the world we live in. Subsequently, it is said that the elementary particles of how many can exist in both wave and particle form.

There are many other complex questions on the subject, but what is important for people like us who study the relationship between quantum physics and the law of attraction are the strange results of some of the experiments that have been conducted. The double-slit experiment, the Copenhagen interpretation and the Schrödinger cat experiment are the main examples.

Double Slit Experiment

The double-slit experiment showed that energy particles could exist in both particle and wave-form, and mere observation by an observer can cause particles to behave differently than before, when they weren't being observed.

Interpretation of Copenhagen and Schrödinger's Cat

The Copenhagen interpretation also states that particles can take on a wave-form or a particle shape depending on the observation. In simple terms, this means that events can change according to the act of observation. Schrödinger was a physicist who suggested the Schrödinger cat experiment to explain his theory better. According to this mental experiment, a (suspected) cat enclosed in a box with a radioactive meter and hydrogen cyanide could be alive and dead at the same time, being both real conditions.

The final result of the experiment depends on the observation made when opening the box to check the cat. According to the Copenhagen interpretation, the act of observation is the decisive factor that determines the reality of the result.

Application of quantum physics to the law of attraction

Our points of view are a type of vitality, and the littlest piece of this vitality is called quantum. We realize that quantum material science expresses that these quanta can exist both as waves and as particles. We also know that the quantum decision to represent itself as waves or particles is dependent on and can be changed depending on the observer's act of observation.

From a deeper understanding, we can conclude that our thought processes can project/create a reality and that reality depends entirely on our act of observation. Or, more simply, let's see what we believe/want to see. We create our reality based on the energy released by our thought processes.

QUANTUM THEORY

Conceptual Revolution

Quantum theory has revolutionized modern physics. It is the part of material science that manages amazing little marvels (nuclear and subatomic occasions). The word 'quantum' means 'the smallest indivisible quantity,' for example, a bright photon or a small measure of angular momentum. A quantum of everything cannot be reduced to lesser quantities. The revolutionary idea of quantum theory is precisely this fact - that events in the observable world occur only in discrete steps from one level of energy to another, without a continuous gradation in between.

Quantum theory emerged from scientists' efforts to explain some measurements of specific physical events such as black body radiation and the photoelectric effect. Classical physics simply could not explain these events. The mathematics of classical physics failed at the atomic and subatomic level, where these events originated. Quantum theory has gradually developed into a highly successful scientific instrument, with which precise measurements with twelve decimal places can be obtained. At the same time, however, he presented us with conceptual puzzles.

Despite its extreme practical precision, quantum theory does not give us a precise idea of what objective reality could be. The question of 'objective reality' seems irrelevant, even absurd from a quantum theorist.

Modern physics

The usual attitude of quantum physicists seems to be that quantum theory shouldn't tell us anything about real objects or real events.

These physicists do not recognize the small-scale objective reality in which human senses do not function. In their view, gross anatomical senses cannot access this reality at this level, since there is no reality here that the human senses can access.

From quantum physicists, science does not deal with 'objective reality' issues, but with coherent measurements and calculations of intertwined processes that allow us some degree of control over our lives. There is absolutely no individual correspondence between the measurements, and there is a unique reality.

Classical physics

Classical physics has always assumed a one-to-one correspondence between physical objects in real-world and measurements that clearly defined these objects. However, quantum theory rejects it as a false fantasy. Irreducible quantum theorists do not accept the existence of a reality separate from measurements at all. Rather, they argue that the measurement process adapts to reality realization. The observer intertwines in what is observed. All science can do is describe the state of the intertwined systems observed by the observer without saying that the observer or the observable have different identities.

Death of sensory understanding

Quantum theory has amputated the human senses from scientific understanding. Physics has not become physical. Science has become hypermathematical in terms of probing, mechanical and measuring devices, leaving the human body sensations in an abandoned wasteland that the new quantum standards consider archaic.

Why should this condition affect us?

I believe that people build successful civilizations by refining gross anatomical sensations that all people share. Common, natural languages, rich in metaphors and analogies, are a means of conveying these refined anatomical sensations. The stability of civilizations could even depend on our successful exchange of metaphors and analogies.

One of the conceptual riddles of quantum theory is the suggestion that all apparent objects and events are inextricably linked, while at the same time requiring rigidly separated energy packets and, in some interpretations, rigidly separated multiple universes that are insensitive to interactions. This creates confusion. If perceived objects and events are unified at the most basic level, humans can only achieve this through generally conceived natural languages that harmonize with special mathematical languages.

The quantum theory in its current form seems contradictory, although it works conceptually. This does not help to unite people and human civilizations to the depth necessary to make great progress.

Aesthetic revolution

I, therefore, suggest that an aesthetic revolution must take place with a scientific revolution and that the modern world is behind schedule. Instead of using a heartless wave function that only offers non-essential probabilities, I wonder if there is an aesthetically pleasing alternative that allows people of all interests to move forward together. Can we better interpret the wave function as measuring a real physical substance in motion? Again, people seem to need such objectivity to understand science.

Flowing worldview

All life as we know it came from the flowing ocean. Perhaps every existence as we know it came from a quantum-superfluid sea. Religions around the world always include water in their creation stories. The surface of the earth consists largely of water, as does the earth's atmosphere and the human body. All indications are visible to inspire a new, modern view of 'objective reality.'

QUANTUM PHYSICS FOR BETTER HEALTH

We are now in the Aquarian era. The era of science. Everything is required by law.

This world has a galactic sun, around which our nearby planetary group pivots.

The primary contrast is in time and extension. Go through a month in each character. Our whole close planetary system will move around the galactic sun in roughly 24,000 years. He goes through 2000 years in each sign.

We, as a people and as a close planetary system, have never been to this piece of the world.

We are in a pristine S.T.E.M. setup (space-time-vitality matter).

S (space) We as a close planetary system are in an extraordinary physical space right now. We have never been here.

T (time) We have no clue how our man-made idea of time is influenced. However, there will be a distinction. I feel that this time, as we are probably aware, is getting quicker.

E (vitality) Einstein reveals to us that E = MC is squared. At this point, we have no idea (less than 50 years after the Age of Aquarius) whether this will apply in the future.

M (Matter) This is the substance of which our physical reality is made. This is what we use to create our thoughts. We have all consumed MATER (parent substance), which has been allowed for the past 200 years. We used it to build the forms and functions of the fish age.

Now we are in an S.T.EM. location in the galaxy called Aquarius.

This new space in our galaxy with its new S.T.E.M. gave us our first two tools. The laws of quantum physics and the internet.

Quantum physics is finally clarifying it. It tells us that all energy in the universe, all information about creation in the spirit of God, exists in an infinite ocean of energy called the quantum ocean. It's all there. It exists in a substance manifested in a timeless/spaceless ocean.

There are divine plans in the quantum ocean for everything on the physical level. There is a divine plan for perfect health.

Thoughts are items. Keep this in mind, and you will draw energies from the quantum ocean into your aura.

We are souls in one body. In reality, we are not physical. We are spiritual beings who are made of the same spiritual material that exists in the quantum ocean.

Since the quantum sea is an unending expanse of reasoning substance, we can interface and communicate with it by thinking.

It is our thoughts that extract the building blocks of our reality from the quantum ocean according to the law of attraction.

As soon as you change your thinking from disease to perfect health, the divine model of perfect health flows into your aura and takes away all the blocked energies that cause disease.

The Frenchman Coue said: 'Every day I am getting better in every way.'

Two magic words open the doors of the quantum ocean. 'I am'.

Whenever you think, speak or write with these two magic words, you draw these actual energies from the quantum ocean into your aura.

Whatever energies, vibrations or thoughts you bring into your aura, they attract the same energies and vibrations from the quantum ocean—Law of attraction. You may like it. What you think fills your aura.

QUANTUM PHYSICS AND YOU

Enter The Quantum Ocean

We have just entered the era of Aquarius. This means that our solar system has turned into a completely different space in the galaxy.

We, as humans on planet Earth, have never been here. It remains to be seen what we can expect from this new space.

At the beginning of a great cycle, there are always some precursors, suggestions, and general rules for what we can expect.

We have occupied this place in space for fifty years. Another 1950 years are still missing. What this new 'era' has already revealed to us is very important for the direction in which we should go.

First of all, the 'Fish Age' we just left is over. Although many of the ideas, institutions and forms of the Piscean era seem strong and surviving, their thorn has been unplugged. You are no longer connected to the power of the Galactic Sun. The galactic sun is driving the Aquarian era.

The age of the fish was known as the age of 'Credo.' It served its purpose because humanity had to build its belief system with authority.

The period of Aquarius is the time of 'I know.' We will have the option to acquire all the data we have to develop actually and advance directly from the source. There will be no requirement for a centerman.

This is made conceivable by one of the primary blessings that made us the 'Period of Aquarius.' These laws will build up rules for the following 1950's.

They just disclose to us that there is a boundless expanse of considerations, shrewd vitality and quantum sea. It is the soul of God.

There is no time, past, present or future—only the HOUR. There is no place. No width, length, or depth only HERE.

In reality, the quantum ocean, the spirit of God, is an infinite point called HERE and NOW. And how we can exist both within this infinite point and outside this infinite point on planet Earth is still a mystery to our finite mind.

Right now, we have to use the laws of quantum physics and the concept of the quantum ocean and the spirit of God, to restore a new reality for us and humanity in general.

Another idea is that the quantum ocean, the spirit of God, responds to our thoughts. Thoughts are things!

We have to do more than just think in the quantum ocean, spirit of God, we have to learn how to get there. How to be there. We must learn to live all day both mentally in the quantum ocean and physically on planet Earth.

We must learn how to apply the laws of quantum physics in the quantum ocean to rebuild our lives and our world.

Your thoughts

At the point when agony and enduring happen upon you, realize that you have brought it into your life. So, face them. Take care of the business! Be a lady! Face the storm they bring and before you know it, the sky turns blue once more. It was only a test, and it made you more grounded.

The time of Aquarius gave us the laws of quantum material science. The laws of quantum material science reveal to us that there is an unbounded expanse of contemplations, savvy vitality. We call this the quantum ocean, the spirit of God. It responds to our thoughts.

Be aware that your every thought draws energy from the quantum ocean. All defective thoughts are thoughts that draw energies from the quantum ocean that are not part of the divine plans created for the guidance and growth of man.

Since man has the power to create together with his thoughts, he has understood how to think outside the box of divine plans. He created his bad schemes. It is the energy in these imperfect man-made patterns that causes all the pain, disease, poverty and misery that surrounds us.

All physical diseases have their origin in the mind of man, not in the quantum ocean, in the mind of God. They are stored there and guided by the quantum ocean, but they do not come from there.

As long as man thinks about these wrong thoughts, he will continue to draw their energy from the quantum ocean into his life.

Do you want to stop all diseases and accidents on the planet? So, stop thinking about it and focus on it! If all men and women stopped thinking and talking about the disease, poverty, and misfortune, they would disappear from physical reality. They would remain inactive as unmanifested energies in the quantum ocean, the spirit of God.

It is our thoughts that trigger pain, disease, poverty and unhappiness in our lives and on our planet. The whole world must be transformed without speaking and thinking about disease, poverty and misfortune. Only thoughts of health, prosperity and happiness should be allowed on television, in the media and in schools.

Doctors and scientists should start looking for health, prosperity and happiness in the world around them. You should start preaching. It is our weakness of will that leaves room for bad habits of thanksgiving that will ruin us in the end.

Start with the kids at school. Teach them to think and expect nothing but health, wealth and happiness. And those will be the energies they will draw from the quantum ocean, from the spirit of God and into their lives.

However, religious leaders and prophets have told us the same thing. Clean your temple, drive out all the demons of your lower nature. What are these demons? They are your wrong thoughts.

QUANTUM PHYSICS - THE DISCOVERY THAT SCIENTIFICALLY DEMOLISHED MATERIALISM

The quantum model of the universe is an attempt to free the big bang from its creationist implications. The proponents of this model are based on observations of quantum physics (subatomic physics). In quantum physics, it can be observed that subatomic particles appear and disappear spontaneously in a vacuum. If some physicists interpret this observation so that matter can arise on a quantum level, it is a property that refers to matter. Some physicists try to explain the emergence of matter from non-existence during the creation of the universe as a property related to matter and to represent it as part of natural laws.

However, this syllogism is definitely out of the question and can in no way explain how the universe was born. William Lane Craig, author of The Big Bang: Theism and Atheism, explains why:

A quantum generation material of the mechanical vacuum is far from the usual idea of a 'vacuum,' meaning nothing. This is not 'nothing,' and therefore, material particles do not arise from nothing.

Matter does not exist in prior quantum physics. What happened is that environmental energy suddenly becomes matter, and then, just as suddenly it becomes energy again. In short, there is no existential condition from nothing, as it is claimed.

As per Isaac Newton, the light was a progression of a substance known as the body. The basis of traditional Newtonian physics - which was accepted until the discovery of quantum physics - was that light consisted entirely of a collection of particles. However, James Clerk Maxwell, a 19th-century physicist, suggested that light demonstrates wave motion. Quantum theory has reconciled this most significant debate in physics.

In 1905, Albert Einstein claimed that light had quantum or small packets of energy. These energy packets were called photons. Although described as particles, photons have been observed to behave in the wave motion proposed by Maxwell in 1860. Therefore, the light was a transition phenomenon between wave and particle (George Gilder), a state that showed a great contradiction in terms of Newtonian physics.

Immediately after Einstein, the German physicist Max Planck examined the light and surprised the whole scientific world by discovering that it was both a wave and a particle. As indicated by this thought, which he proposed under the name of the quantum theory, the vitality was conveyed as hindered and discrete packets, instead of being straight and constant.

In a quantum event, the light showed both particle and wave-like properties. A wave in space accompanied the particle known as a photon. In other words, the light moved through space like a wave but acted as an active particle when it encountered an obstacle. To put it differently, it took the form of energy until it encountered an obstacle. At that point, it took the form of particles as if they were made up of tiny material bodies reminiscent of grains of sand.

According to Planck, this theory has been further developed by scientists such as Albert Einstein, Niels Bohr, Louis de Broglie, Erwin Schrödinger, Werner Heisenberg, Paul Adrian Maurice Dirac, and Wolfgang Pauli. Every one received the Nobel Prize for their discoveries.

Amit Goswami says this of the discovery about the nature of light:

When light is viewed as a wave, it appears to be able to find itself in two (or more) positions simultaneously, as if it passed through the crevices of an umbrella creating a diffraction pattern. However, when we capture it on photographic film, it appears discreetly point by point like a particle beam. So, the light must be both a wave and a particle. It is one of the barriers of ancient physics; a clear description in language. The idea of objectivity is also at stake; does the nature of light or what light is depend on how we view it?

Scientists now no longer believed that matter was made up of inorganic and random particles. Quantum physics had no materialistic meaning because there were unnecessary things like matter.

De Broglie's discovery was extraordinary; in his research, he observed that even subatomic particles exhibited wave-like properties. Even particles like the electron and the proton had wavelengths. In other words, contrary to materialistic belief, there were waves of immaterial energy within the atom, which materialism called simple matter. Just like light, these small particles in the atom sometimes behaved like waves and showed the properties of the particles in others. Contrary to materialistic expectations, simple matter in the atom could be detected at certain times, but disappear on other occasions.

This vital discovery showed that what we imagine as the real world was a shadow. The matter had moved entirely away from the field of physics and was moving towards metaphysics.

Physicist Richard Feynman portrayed this intriguing reality about subatomic particles and light:

Now we know how electrons and light behave. But what can I call it? When I say they act like particles, I make the wrong impression. Even if I say that they act like waves this isn't correct. They behave in their inimitable way, which could technically be called quantum mechanics. A particle doesn't act like a weight that hangs and swings on a spring. It is also not like a miniature representation of the solar system with small planets moving in orbits. Mo more does it look like a cloud or mist surrounding the core. It's like nothing you've ever seen before.

There is at least one simplification. In this regard, electrons behave just like photons, they are both crazy.

The way they behave requires a lot of imagination to appreciate them because we will all describe something different from everything we know. Nobody knows how it is.

All the most famous physicists of the 1920s, from Paul Dirac to Niles Bohr and Albert Einstein to Werner Heisenberg, have tried to explain these results using quantum experiments. Finally, a group of physicists at the Fifth Solvay Physics Conference in Brussels in 1927 - Bohr, Max Born, Paul Dirac, Werner Heisenberg and Wolfgang Pauli - reached an agreement known as the Copenhagen interpretation of quantum mechanics. This name derives from the workplace of the leader of the Bohr group, who suggested that the physical reality proposed by quantum theory is the information we have about a system and the estimates we make based on this information. In his view, these assumptions made in our brain had nothing to do with external reality.

In short, our inner world had nothing to do with the real, outer world, which was the primary interest of Aristotle's physicists to the present day. Physicists have given up on their old ideas about this view and have agreed that quantum understanding is only our knowledge of the physical system.

The material world that we can perceive exists only as information in our brain. In other words, we can never have direct experiences with the matter in the outside world.

Jeffrey M. Schwartz, neuroscientist, and professor of psychiatry at the University of California, described this conclusion from the Copenhagen interpretation:

John Archibald spoke: 'No phenomenon is a phenomenon until it is an observed phenomenon.'

Amit Goswami expanded this result:

Suppose we ask; is the moon there on the off chance that we don't take a gander at it? To the degree that the moon is, at last, a quantum object (made entirely of quantum objects), we have to say no, says physicist David Mermin ...

Perhaps the most important and insidious assumption that we take into account in childhood is that of the material world of objects out there, regardless of who the observers are. There is evidence of this assumption. For example, if we look at the moon, we will find the moon where we expect it to be along its classically calculated trajectory. Of course, we project that the moon is always there in space-time, even when we are not looking. Quantum physics says no. If we don't look, the wave of the moon's possibility is spreading, albeit of a small amount. When we look, the wave stops instantly; therefore, the wave could not be in space-time. It makes more sense to adopt an idealistic metaphysical theory: There is no object in space-time without a conscious subject looking at it.

Of course, this applies to our world of perception. The presence of the moon is apparent in the outside world. In any case, when we take a look, everything we experience is our view of the moon.

Jeffrey M. Schwartz embedded these lines in his book 'The Mind and the Brain' relating to the reality showed by quantum material science:

The job of perception in quantum material science can't be exaggerated. In traditional material science [Newtonian physics], watched frameworks have a presence autonomous of the psyche that observes and inspects them. In quantum physics, however, a physical quantity has real value only through an act of observation.

Schwartz also summarized the opinions of various physicists on this topic:

As Jacob Bronowski wrote in 'The Rise of Man,'

'A goal in the natural sciences was to provide an accurate picture of the material world. One of the achievements of physics in the 20th century was to demonstrate that this goal cannot be achieved.' Heisenberg said that the concept of objective reality was 'so blurry.' In 1958 he wrote that 'the laws of nature that we mathematically formulate in quantum theory no longer deal with the particles themselves, but with our knowledge of elementary particles.' 'It's wrong,' Bohr once said, 'to think that the task of physics is to find out what nature is like. Physics is about what we can say about nature.'

After the most fascinating and sensitive experiments that the human mind has been able to develop over 80 years, there are now no opinions on quantum physics that have been proven decisively and scientifically. Nor can any objection be raised to the conclusions of the experiments carried out. Scientists have tested quantum theory in hundreds of possible ways and have received the Nobel Prize on a number of occasions for their work.

Matter, the most basic concept of Newtonian physics and once considered unconditionally as absolute truth, has been removed. The materialists, supporters of the old belief that matter was the only and last block of existence, were confused by the lack of matter suggested by quantum physics. Now you have to explain all the laws of physics in the field of metaphysics.

The shock this caused to materialists in the early 20th century was far greater than can be expressed in these lines. But quantum physicists Bryce DeWitt and Neill Graham describe it:

'No development of modern science has influenced human thought as profoundly as the advent of quantum theory. Torn from secular thought patterns, the physicists of a generation were forced to face a new metaphysics. The difficulties that caused this reorientation to continue today. Physicists suffered a severe loss; their hold on reality.'

THE QUANTUM DIMENSION

Starting from the work of James Clerk Maxwell in the 19th century, it has generally been concluded that light, electricity and magnetism are variations of the same entity called energy. When physicist Neils Bohr and others began to investigate further, and especially at the subatomic level, it was discovered that all powers have a wavy behavior.

This primarily refers to how a quantum particle changes in one place affect another related particle that is several light-years away. For many scientists, this was a painful fact to accept. Even Einstein, who openly declared that 'imagination is more important than knowledge,' stated that entanglement theory has 'disturbing long-distance effects.' However, its implications may shed light on Professor Giacoma Rizzolatti's discovery of the mirror neuron in the 1990s and the '100 monkey theory' of social change.

Physicist Michio Kaku, Ph.D., and physicist Dean Radin, Ph.D., reported something even more enjoyable. They have publicly documented numerous research projects showing that subatomic energy is compromised when attention is drawn to it. In other words, when we move our consciousness, consciously or preconsciously, onto an object, we transform it. This transformation through the power of observation is one of the most valuable insights of our ability to influence our internal or external environment.

The last of the most essential quantum concepts is what is known as the 'quantum puzzle.' The primary requirement here is that when we study the subatomic realm, we find that there are perceptions and reactions to external effects. This is related to the ideas for quantum observation, and entanglement just mentioned. The idea is that there is also a consciousness at this level, which shows a level of intelligence. If you look at the spiritual implications, science and religion without God, which is generally considered unscientific, are disturbingly connected. Furthermore, this seems to be the explanation behind many of the effects of the morphogenetic field discussed by Rupert Sheldrake, Ph.D., the British biochemist, and plant physiologist, who dared to do valuable scientific research on parapsychology.

What quantum physics seems to tell us is that consciousness permeates all reality and that if we focus our minds, the goal is affected to some extent. This has been confirmed again by quantum physics. And finally, the object of our targeted suggestion and imagination changes the distribution of energy in our brain, body and even in our social structure and physical environment.

THE RELATION BETWEEN WAVES AND PARTICLES

Light behaves like a wave one moment and a particle in the next. This particle, called a photon, was verified by Einstein through his experiments on the photoelectric effect. In this experiment, he concluded that energy packets are released and that this packet is the photon. The tests also confirmed the existence of photons. For example, in an experiment, the light was passed through thin slits like wafers, and on the other side, there was a film. When a photon hits this film, it leaves an image at this point.

Whenever a photon hit the film, it left traces in the film, and this continued indefinitely until a very interesting vibration emerged. Photons repeatedly hit the same areas avoiding other areas of the film. The outcome was a progression of light and dull lines on the film. This was a wave theme, in spite of the fact that it was made by the photons hitting the film.

The molecule had shaped a wave design. Different investigations have checked the wave idea of light. Light can be shone through a crystal and separated into various shading frequencies. For these reasons, we accepted the fact that light behaves both like a wave and like a particle. Experiments will show that it is a particle, and other analyses will show that it is a wave.

The Particle Wave

Light is very different from the other waves we know. Light is nothing but energy, which consists of an electric wave and a magnetic wave. These two waves move perpendicular to each other and perpendicular to the direction of movement. For this reason, light is called an electromagnetic wave. A wave on a lake or in the ocean is the energy that moves through a mass. If you dropped a stone into a pond, you would see a ripple moving outward and away from where the rock entered the water. Now, the motor vitality of the rock was consumed by the lake when the stone hit its surface. This energy then moves across the surface of the pond. It is important not to know that water does not run.

What you see is the wave moving through the water while the water remains in the same place. Only the wave made the journey. This is not only the case with liquid bodies. All forms of matter can absorb energy. Take a spoon and tap lightly on a glass. You will not see a wave passing through the glass, but you will feel it, and part of the energy of the impact will pass through the glass. Now hit the glass a little harder, and this time it will break.

Rigid glass is unable to process the amplitude of the energy you have applied, and it breaks. A superior model is the earthquake. A seismic tremor is just a prompt and abrupt arrival of put away vitality. The pressure builds up over time on a fault line until it is released. When released, energy from this structure is also released. The resulting earthquake is a wave that crosses the earth's crust. All these events I have described are examples of energy that flows through a medium in the form of waves. This is not the case with light.

Light does not need a means of passing through it. The light would prefer that nothing happens, because when it does, it slows down a bit, bends, loses energy in the materials, is reflected in the room, etc. There is practically nothing in the room. I say this mostly because there are huge areas of hydrogen, dust particles, space debris about the size of an asteroid basketball. For all this material that exists in the void of the room, there is still a lot of space out there, and light can pass through that space. This is an essential concept because there is no means by which light can move.

Light is an electromagnetic wave. How does this wave behave, and how is it structured? As I said, light acts both as a particle and as a wave. This says a lot about the light because we have the parts, and it's just to solve the puzzle. When light acts as a particle, we know that a particle must be present. If light acts like a wave, we know there must be a wave.

It would be far-fetched to believe that light behaves in a certain way, depending on how we look at it. The light does not say, 'Now that they're experimenting with me like that, it's better for me to act like a particle.' The light does not decide anything. It is what it is, and we must be able to read the clues and formulate the correct conclusions from those clues.

There are two options used to consider how light can show both the nature of waves and particles. Outlined are some necessary rules for the existence of light as we know it:

1) The presence of electric and magnetic waves is required for the survival of the photon.
2) The speed of light must remain above a certain threshold; otherwise, the wave idea of the light will be pulverized.
3) The photon can't exist without the electromagnetic wave segment.
4) The light will bend in the presence of a strong magnetic field.

It is the presence of both components of the magnetic and electric waves that the photon creates. As long as the light moves at high speed, it shows wavy properties. When the light drops below the speed required to keep the wave intact, the light wave decays into a stream of free photons. This is because magnetic and electric waves are responsible for the existence of the photon. The magnetic component of the light wave detects the presence of magnetic fields and reacts in the presence of this field. This refers to the theory that gravity is a manifestation of magnetism on a macro scale. We observe magnetism on a micro-scale daily. The macroscale includes large mass bodies such as moons, planets, stars, solar systems and galaxies, etc.

It is an amount of space that contains a large number of particles, which together create a strong gravitational field. This gravitational field can't separate power. This doesn't simply incorporate attractive fields.

The attractive fields of a planet or another body assume a job. It additionally contains particles like protons and electrons. A planet is only an ocean of these particles. The attractive wave segment of light recognizes the nearness of these huge pockets of particles. Another chance is that the mix of the two waves makes the photon, and the photon is brought about by a gravimetric power.

The 'hub' speculation opposes the idea that that the photon of a beam of light exists at the hub of the wave. The hub is where the attractive part and the electrical segment meet. This is the main point where these two waves will ever converge in an influx of light. It is this convergence of these two wave segments that lead to the presence of photons. Since this electromagnetic wave is ceaseless, these two segments of the wave consistently stream towards one another and intersect at the nodes.

A light beam is generated, which is both a permanent wave and a continuous photon flow. So, when we experiment to see if the light is a wave, it behaves like a wave. If we decide at the same time to find out if the light is a flow of particles, our experiments will also prove this.

The amplitude of the magnetic and electrical components coincides at the right angles. The photon can exist here. Again, the addition of the two parts leads to the existence of the photon. Since it is a wave, the two components of this wave add up continuously at one point.

So, we can imagine that the photon exists in the amplitude of the two waves. The presence of both components of the electromagnetic wave leads to the photon. This is the second option that I mentioned earlier. When the light rotates at high speed, the two waves continue to connect to create the photon. Rotating the light beam at high speed could create an area within the wave that acts like a particle. In the example above, this range resides in the amplitude of both components of the wave. This has some impressive results. The amplitude and wavelength of the electromagnetic wave directly affect the existing photon type.

This is due to Einstein's energy or the quantum package. The photon, which is the quantum, depends on the type of electromagnetic wave. The longer the wavelength, the less energy must penetrate the photon. The energy available in the light wave is the same amount of energy that will be available for the existence of the photon. This is the essence of the 'particle-wave.' They are not separate entities.

The electromagnetic wave cannot exist without the photon, and the photon cannot exist without the electromagnetic wave. As soon as a ray of light leaves its source, it does it like a wave.

The photon also arises immediately after this wave has left the source. So, if the question were, 'What was the first, the wave or the photon?' The answer would be the wave component, but since the photon is created immediately after, it doesn't matter. The two are almost instantaneous.

WAVE-PARTICLE DUALITY

The wave-particle duality is deeply rooted in the fundamentals of quantum mechanics. In the formalism of the theory, all information about a particle is encoded in its wave function. This complex value function roughly corresponds to the amplitude of a wave at any point in space. This function develops according to Schrödinger's equation. For particles with mass, this condition has arrangements that follow the state of the wave condition. The propagation of these waves leads to wavy phenomena such as interference and diffraction. Massless particles, like photons, have no solutions to Schrödinger's equation, that is, a different wave.

The particle-like conduct is generally apparent because of wonders related to estimation in quantum mechanics. When measuring the position of the particle, the particle is placed in a more local state, as indicated by the uncertainty principle. Seen through this formalism, the measurement of the wave function randomly leads to the fall of the wave function in a pointed event at a certain point. In the case of particles with mass, the probability of detecting the particle in a specific position is equal to the quadratic amplitude of the wave function there. The measurement provides a precisely defined position and is subject to Heisenberg's uncertainty principle.

After the development of quantum field theory, ambiguity has disappeared. The field allows solutions that follow the wave equation and are called 'wave functions'. The term particle is used to identify the irreducible representations of the Lorentz group that are approved by the field. A connection as in a Feynman graph is acknowledged as a computationally advantageous estimation, in which it is known that the outgoing legs are simplifications of the diffusion and the internal lines for a certain order are in an expansion of the interaction in the field. Since the field is not local and quantized, the phenomena's previously considered paradoxes are explained. Within the limits of wave-particle duality, quantum field theory provides the same results.

There are two ways to visualize the behavior of wave particles: according to the standard model and according to De Broglie-Bohr theory.

Below is a representation of the wave-particle duality using de Broglie's theory and Heisenberg's uncertainty principle concerning the position and functions of the spatial wave of the moment for a spin-free particle with a mass in one dimension. These wave functions are Fourier transformations from each other.

The more the function of the position space wave is localized, the more it is probable that the particle with the position coordinates is in this area and consequently the function of the impulse space wave is less localized; therefore the possible components of the impulse the particle might have been more common.

On the contrary, the more localized the function of the spatial wave of the impulse, the more likely it is that the particle with these values of the components of the impulse is in this range. The less localized is the function of the position spatial wave so that the position coordinates that the particle could take are more widespread.

Wave-particle duality is a constant problem in modern physics. Most physicists acknowledge wave-molecule duality as the best clarification for a wide scope of watched wonders. However, it is not without controversy. Alternative views are also presented here. These views are not generally accepted by traditional physics but serve as the basis for valuable discussions within the community.

1. Both the particle and wave view

The pilot wave model, initially created by Louis de Broglie and further created by David Bohm to the theory of shrouded factors, suggests that there is no duality, but that a system has both the properties of the particles and the properties of the wave itself. Time and particles are driven deterministically through the pilot wave fashion, or its 'quantum potential', that directs it towards areas of constructive interference rather than areas of destructive interference. This idea is represented by a significant minority within the physical community.

Are particles really waves? In the first experiments, the diffraction patterns were recorded holistically with the help of a photographic plate that was unable to detect individual particles. As a result, the idea grew that the properties of particles and waves were incompatible or complementary, in the sense that several measuring devices would be needed to observe them. However, this idea was only an unfortunate generalization due to technological limitations. Today it is conceivable to recognize the appearance of single electrons and see the diffraction model as a measurable model comprising of numerous little focuses. Quantum particles are particles, but their behavior is very different from classical physics.

2. Wave view only

Carver Mead, an American researcher and teacher at Caltech recommends supplanting duality with a 'waves in particular' vision. As per analyst David Haddon:

'Mead has cut off the Gordian bunch of quantum complementarity. He guarantees that molecules with their neutrons, protons and electrons are not particles by any means, yet also, influxes of issue.'

As a rough proof of the unique wave nature of light and matter, Mead cites the discovery of ten examples of pure wave phenomena between 1933 and 1996, including the omnipresent laser of CD players, the self-propagating electric currents of superconductors and the Bose-Einstein- Condensation of atoms.

Many-worlds interpretation (MWI) is sometimes represented as pure wave theory, even by its author Hugh Everett, who called MWI 'wave interpretation.'

R. Horodecki's three-wave theory relates the particle to the wave. The theory implies that a voluminous particle is a wave phenomenon intrinsically spatially and temporally extended according to a nonlinear law.

The deterministic collapse theory considers collapse and measurement as two independent physical processes. A division occurs when two wave packets overlap spatially and satisfy a mathematical criterion that depends on the parameters of both wave packets. It is a contraction of the overlap volume. In a measuring device, one of the two wave packets is one of the atomic clusters that make up the device, and the wave packets collapse to the maximum volume of such a cluster. This mimics the effect of a dot particle.

3. Particle view only

In the time of the old quantum theory, William Duane developed a pre-quantum mechanical version of the wave-particle duality and others, including Alfred Landé. Duane explained the X-ray diffraction through a crystal, based solely on its particle appearance. The diversion of the direction of each diffracted photon has been clarified as a result of the incautious transmission quantized by the spatially customary structure of the diffraction of the precious stone.

4. Neither the wave or molecule see

It has been argued that there are never cautious particles or waves, yet only an exchange between them. Therefore, Arthur Eddington generated the name 'wavicle' in 1928 to depict objects, a word that isn't used regularly today. One idea is that zero-dimensional numerical centers can't be viewed. Another reason is that the formal representation of these points, Dirac's delta function, is not physical because it cannot be normalized. Parallel arguments apply to pure wave states. Roger Penrose says:

'These 'position states' are idealized wave functions opposite to the impulse states. While the states of the impulse are distributed infinitely, the positional states are concentrated indefinitely. Neither can be normalized.'

5. Modify the relational approach to wave-particle duality

Relational quantum mechanics was developed from the point of view that the particle detection event established a relationship between the quantized field and the detector. This avoids the ambiguity associated with the application of the Heisenberg uncertainty principle; therefore, there is no wave-particle duality.

Although it is hard to draw a line that isolates the wave-molecule duality from the remainder of quantum mechanics, it is still possible to list some applications of this basic idea.

1. Wave-particle duality is used in electron microscopy, where the small wavelengths associated with the electron can be used to visualize objects much smaller than those visible with visible light.
2. Likewise, neutron diffraction utilizes neutrons with a

frequency of about 0.1 nm, the run of the mill separation of particles in a solid, to determine the structure of solids.
3. Photos can now show this dual nature, which can lead to new opportunities for studying and recording this behavior.

THE BUILDING BLOCKS OF MATTER AND WAVE-PARTICLE DUALITY

One of the big questions of all time is: What are things made of? Most of the answers came from a series of blocks to do it all. Philosophers first studied this, then alchemists and then chemists. It took physicists to solve this problem!

Everything consisted of a combination of four elements; earth, air, fire and water. This had the advantage of having only a few predefined blocks and having connections between properties and content.

The next great model was Mendeleev's periodic table. All matter consisted of atoms with one type of atom per element (e.g. iron, oxygen). Over 100 elements have been identified, a large number of them being basic elements!

Take a piece of gold. Cut it into smaller pieces. Every lump will still be gold. After all, it was thought that it would come to a point where it could no longer be cut. The ancient Greek word 'atomos' (meaning 'uncut') was used to refer to small groups of an element.

However, physicists have discovered that electrons came from atoms. This meant that the atoms had to be something else. The hunt had begun.

In the early 20th century, Rutherford studied J.J.Thompson's plum pudding model. This indicated that negative electrons were kept in a positive 'batter.' He hired some of his students, Geiger and Marsden, to test it.

We like to call some objects tables, chairs or stools. We like to call some things waves; sound waves, water waves or Mexican waves. However, some things have led to long discussions about what they are. Light is an important example.

At the end of the 17th century, Newton believed that light was made up of small particles (bodies), and therefore it was made with the matter, but Huygens believed that light was waving. At that time, Huygens won the debate with a series of experiments that showed how light behaves like a wave; as it could spread through a vacuum.

In 1905 Einstein published an article on a dark phenomenon called 'photoelectric effect.' It had been observed that when the light fell on some charged metals, they lost their charge. However, they were not all metals or all types of light. For example, zinc retains its charge when white light shines on it, but loses it when ultraviolet light (the type used in tanning booths) shines on it. This could not be explained if the light was a wave. Einstein realized that this was proof that Newton was right, and that light was made up of particles. He called them 'how many lights' or photons. The branch of physics called quantum mechanics was born.

But wait a minute, Huygens' experiments have shown that light is a wave. These experiments still work today. So, what's up? Surely, they can't both be, right?

Well, they are. Light appears to be both a wave and a particle. It only behaves differently depending on the circumstances. Sound familiar? Yes, it is like mass-energy and space-time. In this case, the principle is called 'wave-particle duality,' and we say that light is made up of waves.

So, if we were firmly convinced that light is a wave, but it turns out to be made of particles and waves, what about the things we strongly believed were just particles?

In 1906, J.J.Thompson received a Nobel prize for proving that electrons are particles. He had done so by showing that they had quantified mass and charge; they arrived in solid lumps instead of being able to have any number of detached particles. In 1937, his son George Thompson received a Nobel prize for proving that electrons were waves. Today we see that they too are waves, and both Thompsons were right.

WHY MAX PLANCK IS CALLED THE FATHER OF QUANTUM PHYSICS

The father of quantum mechanics is a nickname that is applied to multiple people. Indeed, Max Planck, Werner Heisenberg and Erwin Schrödinger had the same aspirations and the same recognition. With his famous Planck equation, Max Planck unwittingly created the broad field of quantum theory and is considered the 'true but reluctant father' of the modern concept of 'quantum energy,' which underlies all quantum phenomena.

Max Planck was a hypothetical German physicist who is viewed as the author of the quantum theory and one of the most important physicists of the 20th century. At the turn of the century, he realized that light and other electromagnetic waves were emitted in discrete energy packets, which he called 'quanta' or 'quantum' in the singular, and that he could take only certain discrete values (multiples of a certain constant), which are now named 'Planck's Constants'.

Planck's first work dealt with the topic of thermodynamics, an interest he acquired through his studies at Kirchhoff, which he admired very much, and by reading the publications of R. Clausius. He has published papers on entropy, thermoelectricity and the theory of dilute solutions.

At the same time, the problems of radiation processes attracted his attention and showed that they had to be considered electromagnetic. From these studies, he has been led to the problem of energy distribution in the full radiation spectrum. Planck was able to derive the relationship between energy and radiation frequency. In an article published in 1900, he announced his derivation of the relationship; it was based on the revolutionary idea that the energy emitted by a resonator could take only values or certain discrete values. The energy for a frequency resonator v is hv, where h is a universal constant, which is now called Planck's constant.

In 1894, Planck addressed the black body radiation problem, noting that the largest amount of energy emitted by a 'black body' (or a perfect absorber) is more in the middle of the electromagnetic spectrum than in the middle of the ultraviolet region, this suggests classical theory. In particular, he studied how the intensity of the electromagnetic radiation emitted by a black body depends on the frequency of the radiation (for example, the color of the light) and the temperature of the body. After initial frustrations, he obtained the first version of his black body radiation law in 1900. Although he described the spectrum of black bodies observed experimentally, he found that it was not perfect.

Later in 1900, he revised his black body theory to take into account the assumption that electromagnetic energy can only be emitted in a 'quantized' form, so that energy can only be a multiple of an elementary unit $E = hv$ (where h is the Planck Constant, which he had previously introduced in 1899, and v is the frequency of the radiation, although quantization at that time was a purely formal theory in Planck's work, and he never fully understood its radical implications (which had to wait for Albert Einstein's interpretations in 1905). Their discovery was considered as the birth of quantum physics, the greatest intellectual achievement in Planck's career and in recognition of this result he received the Nobel Prize in physics in 1918.

This was not only Planck's most important work but also marked a turning point in the history of physics. The importance of the discovery, with its far-reaching effects on classical physics, was initially not recognized. However, the evidence of its validity gradually became overwhelming as its application took into account many discrepancies between the observed phenomena and classical theory. Among these applications and developments, Einstein's explanation of the photoelectric effect should be mentioned.

LAWS OF QUANTUM PHYSICS

Within a few years, scientists developed a coherent theory of the atom that explained its basic structure and interactions. Decisive for the development of the theory were the new results, which indicate that light and matter have properties of both waves and particles at the atomic and subatomic levels. Theorists had contested the fact that Bohr used an ad hoc hybrid of classical Newtonian dynamics for orbits and some quantum hypothesizes to arrive at the vitality levels of nuclear electrons. The new theory overlooked the way that electrons are particles and treated them like waves. By 1926, physicists had built up the laws of quantum mechanics, additionally called wave mechanics, to clarify nuclear and subatomic marvels.

The duality between the wave and the nature of the light particles was emphasized by the American physicist Arthur Holly Compton in an X-ray diffusion experiment carried out in 1922. Compton sent an x-ray through a target material and observed that a small portion of the beam was deflected to the sides at different angles. He found that scattered X-rays had longer wavelengths than the original ray. The change could only be explained by the assumption that the X-rays scattered by the electrons in the target were as if the X-rays were particles with discrete amounts of energy and momentum. When the X-rays are scattered, part of their momentum is transmitted to the electrons. The force electron takes some vitality from an X-beam, causing the X-beam recurrence move. Both the discrete amount of pulses and the frequency shift of the light scatter are completely contrary to classical electromagnetic theory. Still, they are explained using the quantum formula of Einstein.

In his 1923 doctoral thesis, Louis-Victor de Broglie, a French physicist, suggested that all matter and radiation have properties of both particles and waves. Until the emergence of quantum theory, physicists had speculated that matter was strictly particulate. In his quantum theory of light, Einstein suggested that radiation has properties of both waves and particles. Broglie believed in the symmetry of nature and postulated that even ordinary particles such as electrons could have wave properties. Broglie used the old-fashioned word corpuscle for particles and wrote:

'For both matter and radiation, especially for light, it is necessary to introduce the concept of the corpuscle and the concept of wave simultaneously. As it were, the presence of corpuscles joined by waves must be accepted in all cases.'

Broglie's origination was a propelled one; however, at the time, it had no observational or hypothetical establishment. Austrian physicist Erwin Schrödinger provided the theory.

Schrödinger's wave equation

In 1926, Schrödinger's equation, essentially a mathematical wave equation, established quantum mechanics in diffuse form. To understand how to use a wave equation, it is useful to imagine an analogy with the vibrations of a bell, a violin string or a tympanum. A wave equation controls these vibrations since movement can spread like a wave from one side of the object to the other. Some vibrations in these objects are simple modes that easily get excited and have certain frequencies. For example, the movement of the lowest vibration mode in a head is in phase throughout the head with a uniform pattern around it. The maximum amplitude of the vibration movement occurs in the middle of the eardrum. In more complicated modes, with higher frequencies, the movement on different parts of the vibrating eardrum is out of phase, with one part moving inward and another moving outward.

Schrödinger hypothesized that electrons in an atom should be treated like tympanic waves. The different energy levels of the atoms are identified with the simple vibration modes of the wave equation. The equation is solved to find these modalities. Therefore the energy of an electron is obtained from the frequency of the modality and the quantum formula of Einstein $E = h\nu$. Schrödinger's wave equation provides the same energies as Bohr's original formula, but with a much more precise description of an electron in an atom. The most reduced vitality level of the hydrogen particle, called the ground state, is practically equivalent to development in the least vibration method of the eardrum. In the particle, the electronic rush of the core is uniform every which way, has a top in the molecule and has a similar stage all over. Higher vitality levels in the iota have waves that have more significant movements from the core. Like vibrations in the eardrum, waves have spikes and knots that can form a complex shape. The various waveform shapes are related to the quantum numbers of energy levels, including the quantum numbers for the angular momentum and its orientation.

In the year preceding the creation of his wave theory of Schrödinger, the German physicist Werner Heisenberg published a mathematically equivalent system for describing energy levels and their transitions. In Heisenberg's method, the properties of atoms are described by dispositions of numbers, called matrices, and combining them with special multiplication rules. Today, depending on the application, physicists use both wave functions and matrices. Schrödinger's image is more useful for describing continuous electron distributions since the wave function can be made more easily visible. Matrix methods are most useful for numerical analysis calculations with computers and for systems that can be described using a finite number of states, such as electron spin states.

In 1929, the Norwegian physicist Egil Hylleraas applied Schrödinger's equation with his two electrons to the helium atom. He only got a rough solution, but his energy calculation was quite accurate. With Hylleraas' explanation of the two-electron atom, physicists understood that Schrödinger's equation could be a powerful mathematical tool for describing nature at the atomic level. However, exact solutions could not be obtained.

Antiparticles and electron spin

The English physicist, Paul Dirac, introduced a new electron equation in 1928. Since Schrödinger's equation does not satisfy the principles of relativity, only those phenomena in which particles move much slower than the speed of light can be described. To satisfy the conditions of relativity, Dirac had to postulate that the electron would have a certain form of the wave function with four independent components, some of which describe the spin of the electron. Therefore, Dirac's theory took into consideration the spin properties of the electron from the beginning. The remaining components allowed for further electron states that had not yet been observed. Dirac deciphered them as antiparticles with a charge inverse to that of the electrons. The revelation of the position by the American physicist, Carl David Anderson, in 1932 showed the presence of antiparticles and was a triumph for Dirac's theory.

After Anderson's revelation, subatomic particles could never again be viewed as unchanging. Electrons and positrons can be generated by vacuum if there is an energy source such as high energy X-rays or a collision. They can also annihilate each other and disappear into another form of energy. From this point on, much of the history of subatomic physics has been the history of searching for new types of particles, many of which only exist for a split second after they have been created.

Advances in nuclear and subatomic physics

In the 1920s, Rutherford's discovery of induced radioactivity made further progress in nuclear physics. Bombarding nuclei with light and alpha particles has created new radioactive nuclei. American physicist George Gamow, born in Russia, explained the duration of alpha radioactivity in 1928 using Schrödinger's equation. His clarification utilized a property of quantum mechanics that permits particles to 'burrow' through areas where old-style material science denies it.

structure of the nucleus

The constitution of the nucleus was little known at the time since the only known particles were the electron and the proton. Nuclei have generally been found to be twice as heavy as protons. A coherent theory was impossible until the English physicist James Chadwick discovered the neutron in 1932. He discovered that alpha particles reacted with beryllium nuclei to expel neutral particles with almost the same mass as protons. Almost all nuclear phenomena can be understood using the nucleus of neutrons and protons. Surprisingly, neutrons and protons in the nucleus move largely in the orbitals as if their wave functions were independent of each other. The basic theory based on these orbitals is called the basic shell model. It was introduced in 1948 by Maria Goeppert Mayer from the United States and Johannes Hans Daniel Jensen from western Germany and developed into a complete core theory in the following decades.

The interactions between neutrons and nuclei were studied in the mid-1930s by the American physicist Enrico Fermi and others.

Nuclei easily capture neutrons, which, unlike protons or alpha particles, are not repelled by a positive charge from the nucleus. If a nearby isotope of this atomic mass is more stable, the new nucleus is radioactive, converts the neutron into a proton and assumes the most stable form.

Nuclear fission was discovered in 1938 by German chemists Otto Hahn and Fritz Strassmann as part of experiments initiated and explained by the Austrian physicist Lise Meitner. If cut, a uranium nucleus captures a neutron and gains enough energy to trigger the intrinsic instability of the nucleus, which divides into two lighter nuclei of approximately the same size. The fission process releases multiple neutrons, which can be used to generate further cracks. The first nuclear reactor, a device that allows controlled fission chain reactions, was built at the University of Chicago under the direction of Fermi, and the first autonomous chain reaction was performed in this reactor in 1942. In 1945, American scientists produced the first fission bomb, also known as an atomic bomb, which used uncontrolled fission reactions in the uranium or plutonium of the artificial element. In 1952, American scientists unleashed a fusion reaction with a fission explosion in which hydrogen isotopes were thermally combined to form heavier helium nuclei. This was the first thermonuclear bomb, also known as the H bomb, a weapon that can release hundreds or thousands of times more energy than a fission bomb.

QUANTUM FIELD THEORY AND STANDARD MODEL

Dirac not only proposed the relativistic condition for the electron, he additionally started the relativistic treatment of molecule associations known as quantum field theory. The theory allows for the creation and destruction of particles and requires only the existence of adequate interactions that carry enough energy. Quantum field theory also states that interactions can extend over a distance, only if there is a quantum of particles or fields to bring force. The electromagnetic force that can act over vast distances is carried by the photon, the quantum of light. Since theory allows particles to interact with their field quanta, there have been mathematical difficulties in applying the theory.

The theoretical dead-end was broken by a measurement of the American physicist Willis Eugene Lamb Jr. in 1946 and 1947. Using microwave technology developed in World War II, it showed that the hydrogen spectrum is actually about a tenth percent, unlike Dirac's theoretical framework. Later, the German-born American physicist Polykarp Kusch found a similar anomaly in the electron's magnetic moment size. Lamb's results were announced during a famous Shelter Island conference in the United States in 1947. German-born American physicist Hans Bethe and others acknowledged that the so-called 'Lamb shift' was probably caused by electrons and field quanta, which can be generated by emptiness. Previous math difficulties were overcome by Richard Feynman, Julian Schwinger, and Tomonaga Shin'ichirō, who shared the Nobel Prize in Physics in 1965, and Freeman Dyson, who showed that their different approaches were mathematically identical. It has been discovered that the new theory, called quantum electrodynamics, explains all measurements with very high precision. Quantum electrodynamics provides a complete theory on how electrons behave under electromagnetism.

Similarities between weak force and electromagnetism have been found since the 1960s. Sheldon Glashow, Abdus Salam and Steven Weinberg joined the two forces in the electroweak theory, for which they received the Nobel Prize in Physics in 1979. Despite the photon, three field quanta ought to have other qualities; the W molecule, the Z molecule and the Higgs boson. The W and Z particles were bearers of the feeble power, and the Higgs boson was the transporter of the Higgs field, with the outcome that the W and Z particles were heavy and the photon had a zero mass. The discoveries of W and Z particles in 1983 with correctly predicted masses confirmed the validity of the electroweak theory. A particle that was probably the Higgs boson was finally discovered in 2012.

A total of hundreds of subatomic particles have been discovered since the first unstable particle, the muon, that was identified in cosmic rays in the 1930s. In the 1960s, patterns appeared in the properties and relationships between subatomic particles that led to 'quark theory.' A theoretical framework called a standard model was built from the combination of electroweak theory and quark theory. It contains all known particles and field quanta. In the standard model, there are two major categories of particles, leptons and quarks. Leptons include electrons, muons and neutrinos and, in addition to gravity, interact only with electroweak force.

Quarks are exposed to strong force and combine in various ways to form bound states. Associated quark states, called hadrons, include the neutron and the proton. Three quarks together form a proton, a neutron, or one of the massive hadrons known as baryons. A curd together with an antiquark forms mesons like the pion. The force is so great that quarks outside of hadrons cannot be separated. However, the existence of quarks has been indirectly confirmed in various ways. In the 1967 experiments with high energy electron accelerators were conducted and physicists observed that some of the electrons bombarded the proton targets and were deflected at wide angles. As in the experiment on Rutherford's gold foil, the wide-angle deflection implies that hadrons have an internal structure that contains very small charged objects, probably quarks. To account for quarks and their special properties, physicists developed a new theory of quantum fields known as quantum chromodynamics in the mid-1970s. This theory qualitatively explains the limitation of quarks to hadrons. Physicists believe that theory should explain all aspects of hadrons. However, the mathematical difficulties in dealing with strong interactions in quantum chromodynamics are more severe than in quantum electrodynamics, and rigorous calculations of hadron properties were not possible. However, numerical calculations with larger computers seem to confirm the validity of the theory.

QUANTUM FIELD THEORY

The Quantum field theory (QFT) is the mathematical and theoretical structure for the material science of contemporary, rudimentary particles. In a more informal sense, QFT is the expansion of quantum mechanics (QM), which deals with particles, towards fields, e.g. Systems with an infinite number of degrees of freedom. (See the entry on quantum mechanics.) In recent years, QFT has become a topic further discussed in the philosophy of science, with questions ranging from methodology and semantics to ontology. The QFT, which is taken seriously in its metaphysical implications, seems to give an image of the world that contradicts the central classical ideas of particles and fields and even some characteristics of QM.

What is QFT?

Unlike many other physical theories, there is no official definition of QFT. On the contrary, it is possible to formulate a series of completely different explanations, all with advantages and limitations. One of the reasons for this diversity is the fact that QFT has grown gradually in a very complex way. Another reason is that QFT's interpretation is particularly unclear, so the range of options is also unclear. Perhaps the best and most complete understanding of QFT is obtained by treating its relationship with other physical theories, in particular concerning QM. Still, concerning classical electrodynamics, special relativity (SRT) and solid-state physics or general statistical physics, however, the connection between QFT and these theories is complex and cannot be described step by step.

If we consider QM as the modern theory of a particle (or perhaps very few particles), we can think of QFT as an extension of QM to analyze systems with many particles and, therefore, with a large number of degrees of freedom. In this regard, the transition from QM to QFT is not inevitable, but it is advantageous for pragmatic reasons. However, a general threshold is exceeded when it comes to fields such as the electromagnetic field, which are not only difficult but also impossible to manage in the context of QM. Therefore, the transition from QM to QFT allows the treatment of particles and fields within a uniform theoretical framework. Other than that, focusing on the number of particles or degrees of freedom explains why the famous methods of the renormalization group can be used in both QFT and statistical physics. The explanation is essential that the two orders study frameworks to an enormous degree. A vast number of degrees of opportunity, both because you manage fields like QFT and because you study as far as possible, provide an extremely valuable endeavor in factual material science.

Moreover, the inquiries on the number of particles viable give another motivation to which we should extend QM. Neither QM nor its close relativistic expansion with the Klein-Gordon and Dirac conditions can depict frameworks with a variable number of particles. However, this is essential for a theory that describes scattering processes in which particles of one type are destroyed while others are created.

You get a completely different approach to what QFT is when you focus on your relationship with QM and SRT. It can be said that QFT derives from the good coordination of QM and SRT. To understand the initial problem, it must be recognized that QM is not just a potential conflict with SRT, more precisely, the postulate of the SRT location due to the famous EPR correlations of intertwined quantum systems. There is also an obvious contradiction between QM and SRT on a dynamic level. Schrödinger's equation, i. H. The basic law for the temporal development of the quantum mechanical state function cannot possibly satisfy the relativistic requirement that all physical laws of nature are invariant under Lorentz transformations. The Klein-Gordon and Dirac conditions are coming about because of the quest for relativistic analogs of the Schrödinger condition during the 1920s regarding the necessity of Lorentz invariance. In the end, however, they are not satisfactory since they do not allow the main quantum mechanical descriptions of the fields.

You gain a completely different approach to what QFT is when you focus on your relationship with QM and SRT. It can be said that QFT results from the successful coordination of QM and SRT. To understand the preliminary problem, one must realize that QM is not a potential conflict with SRT, more precisely, the SRT locality postulate due to the famous EPR correlations of entangled quantum systems. There is also an obvious contradiction between QM and SRT at a dynamic level. The Schrödinger equation, i. H. The basic law for the temporal development of the quantum mechanical state function cannot possibly meet the relativistic requirement that all physical laws of nature are invariant under Lorentz transformations. The Klein-Gordon and Dirac equations that resulted from the search for relativistic analogs of the Schrödinger equation in the 1920s respect the requirement of Lorentz invariance. Ultimately, however, they are not satisfactory since they do not allow principal quantum mechanical descriptions of fields.

Luckily, it is genuine for different marvels to disregard the SRT's proposes, specifically when the pertinent speeds are little corresponding to the speed of light and when the dynamic energies of the particles are little contrasted with their mass energies (mc2). In any case, it can never be a fitting system for electromagnetic wonders, since electrodynamics, which incorporates a conspicuous portrayal of the conduct of light, is now relatively invariant and in this manner not good with QM. Dissipating tests are another setting wherein QM comes up short. Since the particles included are frequently quickened nearly to the speed of light, relativistic impacts can never again be disregarded. Consequently, to dissipate QFT you must effectively record examinations.

The Basic Structure of the Conventional Formulation

The Lagrangian Formulation of QFT

The vital advance towards quantum field theory is, in certain regards, practically equivalent to the comparing quantization in quantum mechanics, to be specific by forcing recompense relations, which prompts administrator esteemed quantum fields. The beginning stage is the old-style Lagrangian plan of mechanics, which is an alleged diagnostic definition rather than the standard form of Newtonian mechanics. A summed up thought of force (the conjugate or accepted energy) is characterized by setting $p = \partial L/\partial \dot{q}$, where L is the Lagrange work $L = T - V$ (T is the active vitality and V the potential) and $\dot{q} \equiv dq/dt$. This definition can be persuaded by taking a look at the extraordinary instance of a Lagrange work with a potential V which relies just upon the position so that (utilizing Cartesian directions) $\partial L/\partial \dot{x} = (\partial/\partial \dot{x})(m\dot{x}^2/2) = m\dot{x} = px$. Under these conditions, the summed-up energy agrees with the typical mechanical force. In the old-style Lagrangian field theory, one partner with the given field φ a subsequent field, in particular, the conjugate field.

$$\pi = \partial L/\partial \dot{\varphi}$$

Where L is a Lagrangian thickness, the field φ, and its conjugate field π are the prompt analogs of the authorized arrange q and the summed-up (accepted or conjugate) force p in traditional mechanics of point particles.

While the exchanging connections in QM allude to a quantum object with three degrees of opportunity, so one has a lot of 15 conditions, the exchanging connections in QFT include a boundless number of conditions, for example for every one of the limitless space-time of 4 tuples (x, t) there is another arrangement of exchanging connections. This interminable number of degrees of opportunity encapsulates the field character of QFT.

Interaction

Up to this point, the goal was to develop a free field theory. This not only neglects the interaction with other particles (fields) but is even unrealistic for a free particle since it interacts with the field that is generated. For the description of the interactions, such as scattering in particle collectors, we need some extensions and modifications of the formalism. The direct contact between scattering experiments and QFT is given by scattering or by the matrix S, which contains all the relevant forecast information, e.g. contains cross-sections of scattering. The Hamiltonian interaction is necessary to calculate the S matrix. In turn, the Hamilton operator can be derived from the Lagrangian density using a Legendre transformation.

To discuss interactions, a new representation is introduced, the interaction image, which is an alternative to the images of Schrödinger and Heisenberg. For the image of the interaction, the Hamilton operator, which is the generator of temporal translations, is divided into two parts $H = H0 + Hint$, where H0 describes the free system, that is, without interaction and is absorbed in the definition of the fields. The Hint is the Interaction part of the Hamiltonian or 'Interaction Hamiltonian' for short. The use of the interaction screen is advantageous because the movement equations and, under certain conditions, the switching relationships for the interacting fields are the same as for the free fields. For this reason, it is possible to use different results for the interaction fields determined for the free fields. The central tool for describing the interaction is again the matrix S, which expresses the connection between the In and Out states by specifying the transition amplitudes. In QED, for example, it describes a state | in⟩ a certain configuration of electrons, positrons and photons, d. H. Describes how many of these particles there are and what impulses, spins and polarizations have before the interaction. The matrix S provides the probability that this state changes in a given state, e.g. that a certain counter reacts after the interaction. These probabilities can be verified in experiments.

The canonical formalism of the QFT introduced in the previous section only applies to free fields, since the inclusion of the interaction leads to infinity (see the historical part). Hence, the annoyance theory makes up an enormous piece of most QFT distributions. The significance of troublesome techniques is reasonable when one perceives that they build up direct contact among theory and test. Although the strategies of Bother theory have gotten progressively modern, it is somewhat worrying that disruptive methods could not be avoided in principle. One reason for this discomfort is that Perturbation theory is viewed as a matter of sophisticated craftsmanship rather than an understanding of nature. Accordingly, the body of disruptive methods plays a minor role in QFT's philosophical studies. What is decisive, however, is how the consideration of interactions influences the general framework of QFT. An overview of the Perturbation theory is given in Section 4.1 ('Perturbation theory - philosophy and examples,' by Peskin & Schroeder, 1995).

Effective Field Skills and Recycling

In the 1970s, a system emerged in which the technologies of the standard model of particle physics are seen as effective field theories (EFTs) that have a common quantum field theoretical framework. EFTs only describe phenomena that are relevant in a certain area because the Lagrangian only contains terms that describe particles that are relevant to the respective energy area. EFTs are approximate and change with the energy range under consideration. EFTs are applicable only on a specific energy scale, e.g. H. They only describe phenomena in a certain energy range. The influences of higher energy processes contribute to average values, but cannot be described in detail. This procedure has no serious consequences, as the details of low energy theories are largely decoupled from higher energy processes. Both domains are connected only by altered coupling constants, and the renormalization group describes how the coupling constants depend on energy.

The main idea of the EFT is that the theories, that is, H. in particular, the Lagrangian, on which the energy of the phenomena analyzed depends. Physics changes by adapting to different degrees of energy, e.g. New particles can be generated when a boundary layer of energy passes through. On the energy scale this differentiates QFT from Newton's theory of gravity, for example, where the same law applies to both an apple and a moon. However, the laws of different energy scales are not completely independent of each other. A central aspect of the considerations on this dependence are the consequences of higher energy processes on the low energy scale.

In this context, a new attitude towards renormalization developed in the 1970s, reviving previous ideas, the deviations of which derive from the abandonment of unknown processes of higher energies. Low energy behavior is, therefore, influenced by processes with higher energy. Since higher energies correspond to shorter distances, confidence can be expected from the atomistic point of view. According to the reduction scheme, the separation of the component at the micro-level should determine the processes at the macro level, which is H. Here are the low power patterns. For example, as hydrodynamics shows, technologies at different levels are not closely related to practice, because the applicable law at the macro level can be highly independent of data at the micro-level. For this reason, scripts with computing skills play an important role in EFT discussion. The basic idea of this new renormalization story is that the influences of higher energy processes can be localized in some structural properties that can be determined by adjusting the parameters. 'In this picture, the existence of infinity in quantum field theory is neither a disaster nor an advantage. It is only a reminder of a practical limitation: We do not know what is happening at much smaller distances than we can directly look at.' (Georgi 1989: 456). This new position supports the idea that renormalization is the appropriate response to changing fundamental interactions when QFT is applied to processes on different energy scales. The price to pay is that EFTs are valid only within a limited range and should be viewed as an approximation to better theories on higher energy scales. This raises the important question of whether there is one last fundamental theory in this EFT tower that replaces each other with increasing energies. Some people suspect that this deeper theory could be string theory, e.g. H. a theory that is no longer a field theory.

Cross the Standard Model

The 'model of particle physics' is often used following QFT. However, there is a significant difference. Although the standard model is conceptual with a fixed formula, i. H. Three fundamental forces and many elementary particles, QFT is more than a structure, whose applicability is open. While quantum chromodynamics (or

'QED') is part of the standard model, it is an instance of 'quantum field theory' and not part of the QFT. This section only deals with some particularly important tips that go beyond the standard model, but do not necessarily break down the basic framework of QFT.

Quantum Gravity

The standard model of particle physics includes electromagnetic, weak and strong interactions. The fourth basic force in nature, being, has been subjected to the opposition. The basic problem is that the mass, length and time scales of quantum gravity theories are so extremely small that it is almost impossible to test the various proposals.

The most important surviving versions of quantum gravity theories are canonical quantum gravity, loop theory and string theory. The canonical quantum gravity approaches leave the basic structure of the QFT intact and expand the area of the QFT only through the quantization of gravity. Other approaches attempt to reconcile quantum theory and general relativity not by changing the scope of the QFT, but by changing the QFT itself. String theory, for example, offers a completely new perspective on the most basic building blocks; it does not only include gravitation but also formulates a new theory that describes all four interactions uniformly, in the form of strings.

While quantum gravity theories are very complex and far from classical thinking compared to QM, SRT, and GRT, it is not so difficult to understand why gravity is much more difficult to manage than the other three forces. The weak and strong electromagnetic forces all work in a certain space-time. On the contrary, according to the GRT, gravitation is not a temporal interaction, but gravitational forces are identified with the curvature of space-time itself. So, the quantization of gravitation could be like the quantization of space-time, and it is not at all clear what this means. A controversial suggestion is to deprive spacetime of its fundamental status by showing how it 'appears' in a non-space-time theory. The 'appearance' of spacetime, therefore, means that there are some terms derived in the new theory that have some formal characteristics which are usually associated with spacetime.

String Theory

Marine science is one of the most promising candidates for closing the gap between QFT and general relativity by providing a unified framework for all-natural forces, including gravity. The basic idea of a fiber technique is not to think of particles as basic materials, but rather very thin but expanded fibers on one side. This myth has a critical result that the oceans interact at longer distances and not at some point. This difference between the fiber technique and the standard QFT is important because that is why this technique also adds a very powerful and very difficult QFT to handle.

It is so difficult to reconcile gravity with QFT because the typical length scale of gravitational force is very small, on the Planck scale, so that the theoretical assumption of the quantum field of a point-to-point interaction leads to untreatable infinity. In other words, gravity becomes significant (especially when compared with strong interactions), where QFT is most at risk due to infinite quantities. Extended string interaction means that such infinity can be avoided. Contrary to the entities of standard quantum physics, strings are not characterized by quantum numbers, but only by their geometric and dynamic properties. However, 'macroscopic' strings look like quantum particles with quantum numbers. A basic geometric distinction is between open strings, e.g. H. Strings with two ends and closed strings that are like bracelets. The central dynamic property of strings is their type of excitation, i.e. H. How they oscillate.

The reservations against string theory are mainly due to the lack of testability since apparently there are no empirical consequences that could at least be verified using the methods available so far. There are other quirks here that are hard to swallow. One of them is the fact that fiber science means that space-time has 10, 11 or even 26 dimensions. To explain the occurrence of only four space-time dimensions, string theory assumes that the other dimensions are somehow folded or 'condensed' so that they are no longer visible. An intuitive idea can be obtained by thinking of macaroni, that is a tube, that is, H. A piece of two-dimensional pasta that is rolled up but looks like a remote one-dimensional string.

Despite the problems of string theory, physicists do not give up on this project, also because many believe that string theory is still the best candidate among the numerous alternative proposals to reconcile quantum physics and general relativity, with 'loop quantum gravity' as its strongest rival (see the entry on quantum gravity). As a result, string theory has received some attention within the philosophy of the physical community in recent years. Probably the first philosophical study of string theory is Weingard (2001) in Callender & Huggett (2001), an anthology with other related articles. Dawid (2003) (see other internet sources below) argues that string theory has significant implications for the philosophical debate on realism, that is, it speaks against the plausibility of anti-realistic positions. See also Dawid (2009). Johansson and Matsubara (2011) evaluate string theory from different methodological perspectives and come to conclusions that do not agree with Dawid (2009). The standard introductory monographs on string theory are Polchinski (2000) and Kaku (1999). Greene (1999) is also a hugely popular introduction.

Axiomatic Reformulations of the QFT

Defects in the conventional formulation of QFT.

From the 1930s onwards, the problem of infinity and the potentially heuristic state of QFT's Lagrangian formulation stimulated the search for reformulations in a concise and ultimately axiomatic way. Numerous other aspects have increased the discomfort regarding the standard formulation of QFT. The first is that quantities such as total charge, total energy or total momentum of a field cannot be observed since their measurement should take place in the whole universe. Consequently, quantities relating to infinitely large areas of spacetime should not appear among the observables of the theory as in the standard formulation of QFT. Another problematic feature of the standard QFT is the idea that the QFT relates to the values of fields in space-time points. The net effect of the problem is that a point at a point $\varphi(x)$ is not an operator at a Hilbert point. The physical side of the problem is that it will take an infinite amount of energy to measure one point at a time in space.

A third major problem for the QFT standard has led to rebalancing in the world of regular agents. At the center of mechanical mechanics, Schrödinger, Dirac, Jordan and von Neumann realized that the constants of the Heisenberg matrix and the wave mechanics of Schrödinger are only two (uniformly equivalent) representations of the same underlying abstract structure, i.e. an abstract Hilbert space H and linear operators in this space. In other words, there are only two different ways of representing the same physical reality, and it is possible to alternate between these different representations utilizing a uniform transformation, that is, an operation which is an innocuous rotation of the frame from the analog reference. The representations of a specific algebra or group are sets of mathematical objects such as numbers, rotations or more abstract transformations (e.g. differential operators) together with a binary operation (e.g. addition or multiplication) that combines two elements of algebra or the group, see above that the structure of the algebra or group to be displayed is maintained. This means that the combination of two elements in the display space, e.g. a and b leads to a third element that corresponds to the element that results when combining elements corresponding to a b in the algebra or group shown. In 1931 von Neumann gave detailed evidence (a presumption of stone) that the canonical commutation relations (CCR) for the position coordinates and their conjugate impulse coordinates in the configuration space fix the representation of these two groups of operators in the space of Hilbert up to uniform equivalence (von Neumann uniqueness theorem). This means that the specification of purely algebraic CCRs is sufficient to describe a specific physical system.

In quantum field theory, however, Neumann's uniqueness theorem loses its validity, since it is an infinite number of degrees of freedom. Now you are faced with a multitude of irreducible non-equivalent representations of the RACs, and it is not obvious what it means physically and how to deal with it. Since the disruptive unequal representations of CCRs that occur in QFT are all irreducible, their inequality is not because some are reducible while others are not (a representation is reducible if there is an invariant under-representation, i.e. a subset which alone already represents the RACs). Since the unfair irreducible representations (in short IIR) seem to describe different physical states, it is no longer legitimate to simply choose the most convenient representation, just like choosing the most convenient frame of reference. The seriousness of this problem is not immediately clear, since at first glance it may be possible that all but one IIR are physically irrelevant, e.g. H. mathematical artifacts of a redundant formalism. While this appears to apply to most of the available RIs, there appear to be many irreducible representations of the RACs that are not equivalent and physically relevant.

Algebraic Approaches to QFT

From an algebraic point of view, the algebras of the observables instead of the observables themselves should be seen in some representation as a basic unit in the mathematical description of quantum physics. This will avoid the above problems from the start. In standard QM, the algebraic view concerning C * algebras makes no significant difference compared to the usual Hilbert spatial formulation, since both formalisms are equivalent. This is not true anymore in QFT since the endless number of degrees of opportunity prompts consistently unambiguous and irreducible representations of a C * algebra. Sticking to Hilbert's usual spatial formulation, implicitly implies choosing a particular representation. The term C * algebras, introduced abstractly by Gelfand and Neumark in 1943 and so-called by Segal in 1947, generalizes the term algebra B (H) of all operators bounded on a Hilbert H space, which is also the most important example for a C * algebra. Indeed, it can be shown that each C * algebra is isomorphic concerning an algebra (closed by itself, self-added) of bounded operators on a Hilbert space. The limited (and self-adding) nature of the operators is the reason why C * algebras are considered ideal for the representation of physical observables. The 'C' indicates that it is a complex vector space, and '*' refers to the operation that the element A of an algebra maps to its involution (or the suffix) A *, making the conjugate complex more complex. The numbers are generalized to operators. This involution is necessary to define the crucial norm property of C * algebras, which is central to the demonstration of the above isomorphism claim.

Another point where algebraic formulations are advantageous stems from the fact that two quantum fields are physically equivalent if they generate the same algebras of local observables. Such standard reading space technologies are of the same class called 'Borchers,' meaning that they lead to the same S matrix. As Haag (1996) points out, fields are the only tool to 'manage' well-known observations; the system of observations concerning many spatial areas of the end time. The choice of a particular spatial space is more conventional, that is, as long as it belongs to the same class as Borchers. Therefore, it is more appropriate to plan these algebras as basic bases in QFT rather than quantum fields.

The famous attempt to axiomatize QFT is the Wightman field axiomatic of the early 1950s. Wightman ordered stamps on the algebraic polynomial P (O) from stained fields, that is, collections of products from finite fields to finite spaces O. The main point of this method is the replacement of drawings $x \to \varphi(x)$ by $O \to P(O)$. While the use of unequal field operators makes Wightman's calculation method statistically significant, Kuat's 'Algebraic Field Theory' (AFQT), despite possibly being the most successful attempt to improve QFT axiomatically, only requires operators of freezing. The AFQT was founded in the late 1950s because of Haag's work and rapid progress in cooperation with Araki and Kastler. AFQT itself comes in two parts, namely AFQT (Haag-Araki) and AFQT summary (Haag-Kastler, 1964). A familiar method is to use von Neumann algebras (or W * algebras), a singular C * algebra. The 'abstract' summary refers to the fact that algebras are thus manifested in an unbiased tradition and not through the use of explicit operators briefly in a Hilbert site. In the standard QFT, the CCRs, together with the field equations, can be used for the same purpose, i.e. unethical behavior. The common assumption of these QFT axiomatizations is to avoid the approximate approximations of the standard QFT.

However, when trying to do this in a strictly axiomatic way, only one gets 'fixes' that are not as rich as the standard QFT. As summarized by Haag (1996), 'the algebraic approach [...] has provided us with a framework and a non-technical language.'

The Interpretation of the Particles

Many of the QFT developers are in one of two fields when it comes to prioritizing particles or fields in understanding QFT. While Dirac, later Heisenberg, Feynman and Wheeler opted for the particles, Pauli, Heisenberg, Tomonaga and Schwinger placed the first fields (see Landsman 1996).

The Concept of Particles

It seems almost impossible to talk about the physics of elementary particles or QFT in general without thinking about the particles that are accelerated and dispersed in the collectors. However, it is precisely this interpretation that confronts the more developed contrary arguments. There is still the possibility of saying that our classical concept of a particle is too narrow and that we have to loosen some of its limitations. Because even in the theories of classical corpuscular matter, the concept of an elementary particle is not as problematic as one might expect. For example, if the entire charge of a particle were contracted at one point, an infinite amount of energy would be stored in that particle because the repulsive forces become infinite when two charges of the same sign are brought together. The so-called natural energy of a dot particle is infinite.

Probably the most immediate characteristic of particles is their discretion. This property alone cannot be a sufficient condition to be a particle, since other things are also countable without being a particle, e.g. Money or maxima and minima of the standing wave of a vibrating string. It seems that you also need individuality, i. H. It must be possible to say that this or that particle was counted to explain the fundamental difference between highs and lows in a wave pattern and particles. Teller (1995) discusses a particular determinant of spatial and temporal origin, as other possible components of particle theory compared to classical field and wave theories, and compared to spatial quanta theory, which forms the basis for interpretation that Teller advocates. A key discussion of Teller's thinking can be found in Seibt (2002). There is also a detailed interview on quantity self-assessment in 'same particle' mechanical systems. Since this discussion is primarily QM and not QFT, no further details are given here. Also, contact input counseling on quantity regulation: Identification and recognition.

There is another feature that is usually considered central to the particle concept, namely that particles can be located in space. While it is already clear from classical physics that the requirement of localizability does not have to refer to a punctiform localization, we will see that even the localizability in an arbitrarily large but still finite area can be a strong condition for quantum particles. Bain (2011) argues that the classic terms localizability and countability are inappropriate requirements for particles when one takes into account a relativistic theory like QFT.

Finally, there are some possible components of the particle concept that are expressly opposed to the corresponding (and therefore opposite) characteristics of the field concept. While it is a key feature of a field that is a system with an infinite number of degrees of freedom, the opposite is true for particles. For example, a particle can be designated by specifying the coordinates $x(t)$, which refer, for example, to its center of gravity, which requires impenetrability.

After all, classic particles are massive and impenetrable, again in contrast to (classic) fields.

EINSTEIN'S THEORY OF RELATIVITY

The General Basis Of Relativity

At the age of 25, Albert Einstein (1879-1955) developed his theory of relativity, actually the theory of special relativity, in 1905.

Ten years later, in 1916, after much intellectual effort, he published his theory of general relativity.

Both theories have changed the perception of energy, matter and space. The special theory of relativity deals with bodies and particles that move concerning each other at a continuous and constant speed. In contrast, the general theory of relativity deals with accelerated bodies.

We will only discuss the general theory of relativity here, and leave the special theory of relativity for another article.

What are the Foundations of General Relativity?

The general theory of relativity deals with the gravitational interaction between objects. It manages the principles that oversee the monster universe and the worlds inside.

Everything began when Einstein attempts to communicate, quickened developments corresponding to the gravitational field. He claimed that the acceleration of the 'falling' object could be formulated as the movement of the object in the 'gravitational field' rather because of the gravitational force of the sense of the object, as the classical laws of mechanical physics affirmed.

What was the gravitational field that Einstein had defined? He claimed that the space-time domain was curved. Wherever there is a huge mass, it distorts the space (like making a deep curve in space) on which objects fall if they are close enough.

To understand this clearly, consider the following commonly used example: Suppose that a slightly flat rubber fabric is well stretched. Now place a metal ball on this rubber blanket. The metal sphere creates a curve in the fabric. Another object moving on this fabric, if it is close enough, is captured by this curve and accelerates in the direction of the metal sphere. This is the gravitational field, which consists of a curved space that tries to capture moving objects out there.

Although this example is understandable for a two-dimensional rubber fabric case, it is not as easy to understand for the four-dimensional space-time domain.

However, Einstein spent about ten years of hard mental effort and the use of complicated mathematical principles and other geometries to formulate his new revolutionary theory.

This was a revolution. This postulate undermined the 300-year-old classical Newtonian mechanics that dominated the cosmic movement, which did a great job, and was reflected in incredible space travel and humans landing on the moon.

Contrary to the theory of special relativity, which was easy enough to understand and use even for normal people and whose math was easy enough to support, general relativity theory was indeed a new concept, difficult to digest.

Although the new theory has managed to explain some abnormal orbits of some cosmic objects, and although one of the global scientific experiments of 1919 has shown that light is bent near a large cosmic object, as the equation of general relativity predicted, it still took many years for general relativity to take its place in the scientific world.

The theory of general relativity fully agreed with Newtonian mechanics and proved more accurate when dealing with a huge object, and speed was being handled close to the light. For example, Newtonian mechanics could not explain exactly how gravitational forces work and how they are transmitted. It has also been claimed that the force between objects does not work at any time. For example, if the sun suddenly disappears, the earth will immediately feel this change, and its wake will change immediately. According to the general theory of relativity, however, the disappearance of the sun from the earth is perceived only after about eight minutes, the time when the wave of gravitational change makes its way from the sun to the earth, and only then does the Earth's orbit begin to react to change.

Over the years, more and more evidence of general relativity predictions has been found, and more theories have been developed. Black holes, the expansion/inflation of the universe, the big bang and other amusing terms are based on general relativity.

WHAT IS THE THEORY OF RELATIVITY?

Today's term 'relativity' usually implies Einstein's relativity. When you ask someone what the theory of relativity is, the answer is often that it is a relative movement that Einstein invented. But it wasn't that easy.

Galileo Galilei (1632)

Galileo Galilei first described the 'principle of relativity' in 1632 in his 'Dialogue on the two main systems of the world.' Galileo used the example of a ship sailing at a constant speed on a smooth sea without rocking. Any observer carrying out below deck movement experiments cannot see if the ship is moving or stopped.

This led Galileo to conclude that the laws of physics are the same in any system that moves at a constant speed on a straight line, regardless of its speed or direction. Therefore, there is no absolute movement and no absolute rest.

This was a far-reaching and rather visionary intuition of Galileo, more than two centuries before his time. The principle is sometimes referred to as 'Galilean relativity.'

Sir Isaac Newton (1687)

Galileo's work formed the basis of the three laws of the movement of Sir Isaac Newton, which were published in his monumental work of 1687, today known as 'Principia.' So Galileo's principle of relativity is sometimes, perhaps erroneously, also called 'Newton's theory of relativity.'

Newton's three laws of motion can be quickly communicated as follows:

i. An item very still will, in general, stay very still, and an inconsistent article movement will, in general, stay in steady development except if outside power is applied to it.

ii. The quickening of an item is straightforwardly corresponding to the substance of the power following up on it and contrarily relative to its mass.

iii. There are an equivalent and inverse response for each activity.

However, significant these laws are, they have, in some sense, declined backward since the Galilean theory of relativity. Newton declared that his laws were 'absolute space,' which means that there had to be absolute movement and absolute calm. This was caused in part by the vision of time in which light in space moves at a constant speed through an invisible medium called 'ether.'

Albert Einstein (1905)

In terms of world time, Einstein differed from Galileo and Newton. In his pioneering work of 1905, 'On the electrodynamics of moving bodies,' he renounced world time and postulated that every picture of inertia has its time.

Einstein repeated Galileo's principle of relativity, in which the laws of physics are the same in every inertial frame, regardless of their speed or direction. Einstein's knowledge of 1905 expanded Galilee's theory of relativity into a decisive aspect: The speed of light in free space (symbol c) is the same in every inert reference system, regardless of its movement.

By chance, Galileo likewise felt that the speed of light was the equivalent in each inertial edge, but this is only because he claimed that the speed of light was almost infinite. If you took Einstein's special relativity equations and set the speed of light on infinity, you would get Galileo's relativity principle. In a sense, Galileo and Einstein were closer to their principles of movement than Newton and Einstein.

Newton (like Einstein) knew that light has a finite and measurable speed. However, Newton (erroneously) believed that the speed of light is constant only in absolute space, which means it must have different values than moving observers. If Newton had been right, the speed of light would have been different in different directions within a moving inertial structure. This has been excluded from experiments.

Einstein proposed that the overall idea of movement and time directs that all structures of latency measure a similar consistent speed of light every which way. Examinations have affirmed this view. Einstein's uncommon relativity standards can be quickly outlined as follows:

1. There is no noticeable outright space or supreme development.
2. There is no all-inclusive time. Each edge of idleness has now is the ideal time.
3. The laws of material science are equivalent in every aspect of idleness.
4. The deliberate speed of light is autonomous of relative development.

These four standards infer that time and separation estimations contrast in various inertial structures. When two observers move relative to each other, they generally disagree on the distance between two objects and also on the time it takes for light to travel that distance. This is the essence of Einstein's theory of special relativity.

Gravity: Newton (1687) - Einstein (1916)

Einstein's theory of general relativity of 1916 added gravity to his particular theory and substantially replaced Newton's theory of universal gravitation, as described in his 'Principia' in 1687. If the gravitational field is weak and the speeds are very low compared to speed, Newton's theory of gravity is accurate enough for all practical purposes.

Einstein's theory of relativity becomes a prerequisite only if the speeds are significant fractions of the speed of light and / or the gravitational fields are a thousand times stronger than what we experience here on Earth. Such conditions are observed near neutron stars and black holes, but that's another story.

SIX THINGS EVERYONE SHOULD KNOW ABOUT QUANTUM PHYSICS

Quantum physics is usually intimidating from the start.

It's a little strange and may seem counterintuitive even to physicists who face it every day. But it is not incomprehensible. When you read about quantum physics, there are six key concepts you should consider. If you do, quantum physics will be much easier to understand.

Everything consists of waves; even the particles

There are many places where this type of discussion can begin; everything in the universe has both the particle and the wave nature at the same time. There is a line in Greg Bear's fantasy ideology (The Infinity Concerto and The Serpent Mage) in which a character describing the basics of magic says, 'Everything is a wave without sway, no distance.' I've enjoyed that as an idyllic depiction of quantum material science, all things considered, everything known to humanity has a wave nature.

Everything known to man has a molecular nature. This may sound completely insane, but it is an experimental fact that has been worked out through a surprisingly familiar process.

Of course, describing real objects both as particles and as waves is inaccurate. In reality, the purposes described by quantum physics are neither particles nor waves. Still, a third category that has some wave properties (a characteristic frequency and wavelength, some distributed in space) and some properties of the particles (they are generally countable and they can) must be located together to some extent. This leads to a lively debate within the physics community about whether or not to talk about light as particles in physics courses. Not because there is controversy about whether the light has a particle nature, but because naming photons as 'particles' rather than 'excitation of a quantum field' could lead to some misunderstanding by students. I tend to disagree, as many of the same concerns about labeling electrons as particles could be raised, but this is a reliable source for blogs.

This 'entryway number three' nature of quantum objects is reflected in the occasionally mistaken language for which physicists talk about quantum wonders. The Higgs boson was discovered as a particle in the Large Hadron Collider, but you will also hear physicists speak of the 'Higgs field as a delocalized thing that fills the entire room. This is because in certain circumstances, such as in-collider experiments, for example, it is more convenient to discuss Higgs' field suggestions for emphasizing particle-like properties. In other circumstances, such as the general discussion of why some particles have mass, it is more convenient to discuss physics in terms of interactions with a quantum field that fills the universe. It is just another language that describes the same mathematical object.

It's in that spot in the name, 'quantum' originates from the Latin for 'how much' and mirrors the way that quantum models consistently contain something that shows up in discrete amounts. The energy contained in a quantum field arrives in integer multiples of fundamental energy. As for light, this is related to the frequency and wavelength of light; high-frequency shortwave light has large characteristic energy, while low-frequency longwave light has small characteristic energy.

In both cases, however, the total energy contained in a particular light field is an integral multiple of that energy - 1, 2, 14, 137 times - never a strange fraction like one and a half, π or the square root of two. This property is also evident in the discrete energy levels of the atoms and the energy bands of solid bodies; some energy values are allowed, others are not. Atomic clocks work due to the attentiveness of quantum physics and utilize the recurrence of light connected with progress between two permitted states in cesium to keep time at a level that requires the much-talked-about 'second bounce.'

Ultra-precise spectroscopy can also be used to search for things like dark matter and is part of the motivation of an institute for low-energy basic physics.

This isn't always obvious, even some things that are quantum, like blackbody radiation, seem to imply continuous distributions. But there is always a kind of granularity in the underlying reality when dealing with math, and that is a major piece of what prompts the franticness of theory.

Quantum Physics is Probabilistic

One of the most amazing and (in any event verifiably) most disputable parts of quantum physics is that it is difficult to anticipate with assurance the consequence of a solitary trial on a quantum system. At the point when physicists anticipate the result of an investigation, the expectation consistently appears as a likelihood of finding every one of the potential outcomes, and the examinations among theory and test consistently incorporate the finish of likelihood appropriations from many rehashed tests.

The numerical portrayal of a quantum system normally appears as a 'wave work,' which is commonly spoken to in the conditions by the

Greek letter psi: Ψ. There is a lot of debate about what exactly this wave function is. It is divided into two main fields: those that consider the wave function as a real physical thing (the slang term for this is 'ontic' theories that make a funny person doubt their proponents) and those who consider the wave function simply as an expression of our knowledge (or lack thereof) of the underlying state of a particular quantum object ('epistemic' theories).

In both classes of basic models, the probability of finding a result is not given directly by the wave function, but by the square of the wave function (at least vaguely; the wave function is a complex mathematical object i.e. it is imaginary numbers such as the square root of the negative one and the operation to obtain the probability is a little more complicated, but 'square of the wave function' is sufficient to obtain the basic idea). This is known as the 'Born Rule' after the German physicist Max Born, who first proposed it (in a footnote from a 1926 article) and affected some people as a bad ad-hoc addition. Parts of the quantum foundation community are actively trying to find a way to derive the Born rule from a more fundamental principle. To date, none of them have been completely successful, yet, they produce a ton of fascinating science.

This is additionally the part of the theory that prompts the nearness of particles in different states all the while. All we can predict is a probability, and before a measurement that determines a certain result, the system to be measured is in an indeterminate state, which is mathematically mapped to an overlap of all possibilities with different probabilities. Whether you consider it the system that is actually in all states at the same time, or just an unknown state, largely depends on your feelings about ontic and epistemic models, although both are subject to the limits of the next item on the list:

Quantum Physics Is Not Local

Einstein's last major contribution to physics was not generally recognized as such, mainly because he was wrong. In a 1935 article with his younger colleagues Boris Podolsky and Nathan Rosen (the 'EPR document'), Einstein provided a clear mathematical statement about something that had been bothering him for some time, an idea we now call 'entanglement.'

The EPR article argued that quantum physics allowed for the existence of systems in which measurements at distant positions could be correlated in a way that indicated that the result of one was determined by the other. They argued that this meant that the results of the measurement had to be determined in advance by a common factor since the alternative would require that the result of one measurement be transferred to the position of the other at speed greater than the speed of light. Therefore, quantum mechanics must be incomplete, a simple approximation of a deeper theory (a theory of 'local hidden variables' in which the results of a given measurement did not depend on anything further from the measurement site of a signal at a speed of light ('local'). Still, they are determined by a factor common to both systems in an intertwined pair (the 'hidden variable').

This was considered a strange footnote for about thirty years as there seemed no way to test it, but in the mid-1960s, Irish physicist John Bell elaborated on the implications of the EPR document. Chime indicated that conditions could be found in which quantum mechanics anticipate relationships between remote estimations more grounded than any conceivable theory of the sort favored by E, P, and R. This was tentatively tried by John Clauser and in the mid-1970s. By and large, that various Alain Aspect's trials in the mid-1980s indisputably indicated that a nearby theory of concealed factors couldn't clarify these interlaced frameworks.

The recognized way to deal with understanding this outcome is to state that quantum mechanics can't. The results of measurements in a given position can depend on the properties of distant objects in a way that cannot be explained by the signals, which move the speed of the light.

However, this does not allow information to be sent at speeds faster than the speed of light, although there have been numerous attempts to find a way to use quantum non-location to do it. Their refutation proved to be a surprisingly productive company. For more information, see David Kaisers 'How Hippies Saved Physics.' The non-locality of the quantum is also central to the problem of information on the evaporation of black holes and the controversy surrounding the 'firewall,' which has caused a lot of activity in recent times.

Some radical ideas imply a mathematical connection between the intertwined particles and the wormholes described in the EPR document.

Quantum Physics is (mainly) Very Small

Quantum physics has a reputation for being strange because its predictions dramatically contradict our daily experience (at least for humans - the imagination of my book is that it doesn't seem so strange to dogs). This is because the effects involved decrease as the objects get bigger.

If you want to see clear quantum behavior, you want the particles to behave like waves and the wavelength to decrease as the moment increases. The wavelength of a macroscopic object such as a dog walking in space is so ridiculously small that if everything expands so that a single atom in space has the dimensions of the entire solar system, the wavelength of the dog is approximately the size of a single h.a atom in this solar system.

This means that quantum phenomena are largely confined to the scale of fundamental atoms and particles, with masses and speeds small enough to make the wavelengths large enough to be observed directly. In many sectors, however, active efforts are being made to increase the size of systems with larger quantum effects. I wrote on the blog a series of experiments by Markus Arndt's group that show wavy behaviors in ever-larger molecules, and there are many groups in 'Optomechanical cavities' that try to slow down the movement of silicon blocks towards the light Point where the discrete quantum nature of movement would become clear.

Quantum Physics is not Magical

The previous point naturally leads to this, strange as it may seem, quantum physics is not magical. The things he predicts are strange by the standards of everyday physics but are severely limited by well-understood mathematical rules and principles.

So, if someone approaches you with a 'quantum' idea that seems too good to be true - free energy, mystical healing powers, impossible space propulsion - it is almost certain that it is so. This doesn't imply that we can't utilize quantum physics to do astounding things - you can find really interesting physics in earth technology - but these things remain within the limits of the laws of thermodynamics and basic common sense.

RELATIVITY, QUANTUM PHYSICS, AND STRING THEORY

Einstein's general theory of relativity and the theory of quantum mechanics are two theories of physics widely accepted and experimentally verified. There is only one problem. They contradict each other!

The general theory of relativity deals with physics on the macroscale of huge gravitational fields that distort the spacetime continuum. Quantum mechanics is useful for describing particle physics and the interactions between matter and energy on extremely small scales.

The general theory of relativity describes a continuum of spacetime, but quantum physics describes spacetime distortions in 'quantum foam.' Therefore, these well-proven theories are mutually exclusive.

The general theory of relativity is used to describe large-scale gravitational fields and, therefore, rarely collides with quantum mechanics in research. An exception is black holes, in which gravity is so strong that a strong gravitational field collapses at the micro-level.

When studying black holes, the equations of general relativity are combined with those of quantum mechanics and give meaningless answers.

To resolve this conflict, the new theory, still experimentally untested, called string theory was further developed.

A sketch of string theory states that subatomic particles such as electrons, are not mere points, but oscillating strings. This assumption eliminates quantum foam through a bizarre series of steps.

The string theory additionally contains the dubious thought that the universe is comprised of nine or ten spatial measurements, notwithstanding the transient measurement. These extra measurements are called concealed measurements.

With this advancing theory, a few researchers have accidentally taken advantage of the lucky break to plan a theory of cosmology that wipes out the requirement for God as the maker. Paul J. Steinhardt and Neil G. Turok built up a theory that professes to demonstrate that the universe has consistently existed and consequently was not made.

The complicated theory of Steinhardt and Turok, which I will not describe here, does not overcome a fundamental error. The theory ignores the second law of thermodynamics, also called the entropy law.

The law of entropy is firmly anchored in physics and states that any isolated system must tend to become more disordered and randomized over time. So, if the universe were infinitely old, it would be infinitely randomized and disordered and would suffer heat death billions of years ago.

Scientists who deny creation will sometimes try to guess or ignore the second law, but no self-respecting physicist will openly deny the second law. In curves, they must surrender and admit that the second law of thermodynamics is effective even in extreme circumstances.

Strange as it may seem, the way non-believing scientists avoid conflict with the second law is to assume that the universe itself was created. In other words, they believe in a spontaneous generation where something came out of nowhere for no reason.

This denial of causality was first proposed in the 18th century by the Scottish philosopher David Hume, who was one of the most daring skeptics of the enlightenment.

Scientists, including many 'big bang' theorists, suggest that the universe emerged from nowhere to avoid a conflict with the second law of thermodynamics. Still, they fall into a much larger hole. This hole is 'spontaneous generation.'

Spontaneous generation is absurd. For the universe to create something, it must first exist. For the universe to create itself, it must exist and not exist at the same time and therefore start itself. The logical contradiction is obvious and fatal for spontaneous generation.

Given the implications of the second law of thermodynamics, we must conclude that if something exists now, it means that something has always existed.

BASICS ON ANGULAR MOMENTUM ON A QUANTUM LEVEL

In quantum mechanics, the precise force administrator is one of a few related administrators similar to the traditional rakish energy. The precise force administrator assumes a focal job in the theory of nuclear material science and other quantum issues with rotational evenness. In both traditional and quantum mechanical frameworks, rakish force (together with direct force and vitality) is one of the three fundamental properties of development.

There are several operators of the angular momentum; total angular momentum (usually indicated by J), orbital angular momentum (usually indicated by L) and angular moment of rotation (in short, rotation, usually indicated by S). The term operator of the angular momentum can, confusingly, refer to the total orbital angular momentum. The whole angular momentum is always preserved.

Law of Conservation of Angular Momentum

If a skater spins on her skates, she will go faster if she forms a circle in a smaller radius. This is because a variable called angular momentum (AM, and Mom) is maintained. The equation gives the angular momentum = I * w * w. Angular velocity is a vector quantity, which implies that the localized direction of movement is important for circular motion.

If a skater suddenly reduces his range of motion, his moment of inertia decreases. The speed increases to maintain the angular momentum. In other words, if it runs in a smaller circle, its speed increases.

Let's take the case when traveling in a larger circle than before. The angular momentum is also maintained in this case. In this case, the moment of inertia increases. To preserve the entire AM, its speed decreases.

This law is called 'Ang Mom Conservation Law' and applies to all objects that rotate in a circular motion.

An analogy with the same law is the law of conservation of momentum, which applies to movements in a straight line. In this case, the impulse of the colliding body system is preserved. The impulse before the collision is, therefore, $MV + M1V1$, where M and M1 are two different masses traveling at different speeds. In this way, the overall impulse of the system is preserved. AM and linear impulse are vectors that imply that the size depends on the direction.

Angular momentum plays a central role in both classical and quantum mechanics.

In classical mechanics, all isolated systems receive angular momentum, as well as energy and linear moment. This fact significantly reduces the work involved in the calculation, trajectories of planets, rotation of rigid bodies and much more.

Likewise, in quantum mechanics, angular momentum plays a central role in the structure of atoms and other quantum problems involving rotation symmetry.

Like other observable quantities, the angular momentum in QM is described by an operator.

In effect, this is a vector operator, is similar to the impulse operator. In short, unlike the linear momentum operator, we see the three components of the angle and the impulse operator does not swing.

There are several operators of the angular momentum in QM: the total angular momentum, the orbital angular momentum (usually indicated by L ~), and the intrinsic angular momentum or spin (indicated by S ~). The latter (Iap) does not have a classic analog.

Confusingly, the term 'angular momentum' can refer to the total angle, pulse or angular momentum of the orbit.

It is possible to adopt the classical definition of the orbital angular momentum L ~ = ~ r × ~ p, directly to QM by reinterpreting ~ r and ~ p as operators associated with position e , the linear moment.

Angular Momentum Quantum Numbers

There are a lot of precise force quantum numbers related to the vitality conditions of the molecule. As far as old-style material science, rakish force is a property of a body that is in the circle or is pivoting about its hub. It relies upon the rakish speed and appropriation of mass around the hub of upheaval or pivot and is a vector amount with the course of the precise energy along with the turn hub. As opposed to traditional material science, where an electron's circle can expect a persistent arrangement of qualities, the quantum mechanical precise force is quantized.

Moreover, it can't be indicated precisely along each of the three tomahawks all the while. For the most part, the rakish force is indicated along a pivot known as the quantization hub, and the greatness of the precise energy is constrained to the quantum esteems square root of $\sqrt{l(l + 1)}$ ($\hbar$), in which l is a whole number. The number l, called the orbital quantum number, must be not exactly the primary quantum number n, which relates to a 'shell' of electrons. Accordingly, l isolates each shell into n subshells comprising of all electrons of a similar head and orbital quantum numbers.

There is an attractive quantum number additionally connected with the rakish force of the quantum state. For a given orbital force quantum number l, there are 2l + 1 vital, attractive quantum numbers ml extending from −l to l, which confine the division of the all-out rakish energy along with the quantization hub with the goal that they are constrained to the qualities $ml\hbar$. This marvel is known as space quantization and was first exhibited by two German physicists, Otto Stern and Walther Gerlach.

Basic particles, for example, the electron and the proton additionally, have a steady, natural, precise force, notwithstanding the rakish orbital energy. The electron carries on like a turning top, with its natural, precise energy of greatness s = square root of $\sqrt{(1/2)(1/2 + 1)}$ ($\hbar$), with considerable qualities along with the quantization pivot of msh = $\pm(1/2)\hbar$. There is no old-style material science simple for this purported turn precise energy. In essence, the rakish natural force of an electron doesn't require a limited (nonzero) span. However, old-style physical science requests that a molecule with nonzero precise energy must have a nonzero sweep. Electron-crash concentrates with high-vitality quickening agents show that the electron demonstrations like a point molecule down to a size of 10−15 centimeter, one-hundredth of the range of a proton.

The four quantum numbers n, l, ml, and ms indicate the condition of a solitary electron in an iota totally and extraordinarily; each arrangement of numbers assigns a particular wave work (i.e. quantum condition) of the hydrogen particle. Quantum mechanics determines how all out rakish force is developed from the part precise momenta. The part precise momenta act as vectors to give the all-out rakish force of the molecule. Another quantum number, j, speaking to a mix of the orbital precise energy quantum number l, and the turn rakish force quantum number s can include just discrete qualities inside a molecule: j can take on positive qualities just between $l + s$ and $|l - s|$ in the whole number advances. Since s is 1/2 for the single electron, j is 1/2 for l = 0 states, j = 1/2 or 3/2 for l = 1 states, j = 3/2 or 5/2 for l = 2 states, etc. The size of the complete, precise energy of the molecule can be communicated in a similar structure concerning the orbital and turn momenta: Square root of $\sqrt{j(j+1)}$ ($\hbar$) gives the extent of the all-out rakish force; the part of rakish force along the quantization hub is $m_j\hbar$, where mj can have any an incentive among +j and −j in the whole number advances. An elective depiction of the quantum state can be given as far as the quantum numbers n, l, j, and mj.

THE VERY ESSENCE OF SPACE-TIME AND GRAVITY

The pulling force that holds all matter together is called gravity. If the matter increases, the gravitational force also increases. The mass measures the amount of matter in something. When things are huge, they exert more gravitational force. An excellent example of gravitational force is the force exerted by the earth. When we walk, the earth pulls us there, and, on the contrary, we pull the earth towards us. However, since the earth is huge, we cannot move it. The power of the earth is much higher and strong enough to drop a person face down.

Gravity depends on the mass and also on distance. This is exactly why we remain on Earth and are not attracted to space by the Sun, which has a gravitational force that is many times stronger than that of Earth.

All About Spacetime and its Bases

Things begin and happen in space. Time is seen as the time when these things happen. It is necessary to combine these two elements to understand the universe.

Another fundamental fact is that even if the light is moving at 299,792,458 meters per second and you are on the move, you may notice that time is moving faster, which is not the case.

The Anti-Gravity Wheel

A 19 kg wheel is raised above the head with one hand while the wheel rotates at a few thousand rpm. The wheel seemed light as a feather at the time, and this was due to the gyroscopic precession.

The weight of the object creates torque, and therefore the wheel feels light when it turns. It can be seen that the pair vector increases the angular momentum in the same direction as the pair. If at the beginning there is no angular momentum, the new moment oscillates in the direction of the couple. If there was an angular momentum at the beginning, the direction is changed in the direction of the angular momentum, causing the precession.

Stephen Hawking's Journey Through Time

Stephen Hawking examined the actual requirements for the concept of travel time. He gave detailed instructions on how to build a time travel machine. He says the requirements are very simple because a wormhole, rocket or large hadron collector is needed to build the machine.

The premise is rooted in Einstein's theory of relativity, and Hawking believes that because time moves faster in certain places, you can use it to move and travel to the future, but you cannot travel to the past.

CONCLUSION

Like all sciences, quantum physics is an endeavor to characterize reality. What's more, despite the fact that quantum material science is, by definition, a science, its astonishing information has obscured the line between old style science and its age-old philosophy, religion and mysticism.

The quantum line of the investigation did not define reality but redefined reality. It is as if we had started turning the car, and we found somewhere on the track that the car is not a car, but a mobile device. Deepak Chopra said something like, 'The universe is not only stranger than we think, it is also stranger than we can think.' This is undoubtedly a true statement when trying to understand quantum physics.

However, the basic ideas of quantum physics are beneficial if you keep them on an intuitive level. Consider these three basic principles:

1. A central tenet of quantum physics is that there is an undulating ocean of possibilities on the necessary levels of matter, which physicists have described as a 'unified field.' All matter and form derive from this indefinite and nebulous network of possibilities and potentialities.

2. Experiments with subatomic particles have shown that physical space and time are irrelevant on this almost unimaginable level of matter. Particles can not only present themselves as waves or particles, but they can also exist in two places simultaneously (called 'overlap'). Quantum science has also provided evidence that past experiences are not only about the present (the fact that we expected), but that future events can also bring about changes in current reality.

This is because the sacred notions of space and time become basic builds of the psyche, imagined to make structure and request even with a disorderly universe.

3. Concerning the 'constructs of the mind that were invented to create form and order in the face of a chaotic universe,' ... physicists already learned in 1927 the meaning, and the effect, of the observer about a given reality. Heisenberg's famous uncertainty principle emerged from the awareness that a particle does not change only when it is observed, but when the viewer changes the way he thinks/perceives an inevitable reality and also when the given reality changes with the change.

So, there is no absolute, objective reality. EVERYTHING is subjective.

What would it mean if this statement were true? Some people may raise their hands and say, "Let the chaos begin" or "Such an idea would favor a culture of egomaniacs" or "How can we then think of the ultimate realities like God, justice, and truth?"

If you pause and consider the question honestly, it becomes increasingly clear that our perception of reality is already subjective. A simple test is based on the fact that if twenty-five of us went to the same party, there would be many different experiences with the same event.

Regardless, the idea that anything and everything is subjective is quite radical. It is not surprising that eighty years after the first 'quantum' discoveries, we are still in the grip of a worldview that insists on concrete and true reality. Unfortunately, there is no place for creativity, awareness, spirituality or unity in this almost universally accepted 'universe of watches.' And this is a very convenient place for those who ignore (or worse, distrust) creativity, awareness, spirituality or unity.

Fortunately, due to the revelations of quantum physics, we must all learn to think differently. The beginning of the 21st century offers the opportunity for a new paradise, new earth, and the use of the already worn-out expression, a new paradigm for looking at our world. A new reality is truly at our door.

The number l, called the orbital quantum number, must be not actually the essential quantum number n, which identifies with a shell of electrons. Here are some things that happen in a universe where the observer (you, me and every other conscious being) has an impact on what constitutes reality:

1. The experience becomes interactive. We are no longer passers-by in a static landscape that floats in and out in different situations without changing anything.
2. The experience becomes flexible. A version of a reality that flows and changes dynamically is put before our eyes.
3. The experience becomes creative. Seeing yourself as an integral part of every experience opens the locks for personal will and creativity. How can we use our attention, intention, hopes, and dreams in such a version of reality?
4. The experience becomes more interesting. We recognize that people and animals have conscious awareness. Many of us think of plants, trees and other living things to have similar knowledge. But confronting the truth that even the most essential elements (like an electron) have consciousness is a surprising revelation for most people
5. The proven effect of the observer dispels the illusion of separation. And here he is, down and dirty:

All things within the unified field are intertwined, and the existence of a thing/observer affects the whole system. It is up to the observer to determine whether the commitment, and the resulting impact, is intentional.

Be a conscious observer; create a new reality with thoughts and intentions... why is it important, and how does it work? It is only essential if you want to change something in your life. It works because new thoughts and ideas turn the biochemistry of the brain and eventually replace the physical body as a whole.

As a thought moves through the brain, self-induced biochemists (peptides) enter the bloodstream and generate feelings, emotions, impulses, and all kinds of conscious and unconscious reactions. These reactions are ultimately related to the thought that triggered this particular waterfall. If the peptides get stuck anywhere in the body, the character of that cell also changes. So, when we think we're sad, broken or in love, our body chemically corresponds to thought, and we become literally and physically 'sad', 'broken' or 'in love.'

On the contrary, any change you desire can begin with a change in the way you think. Does this mean that we can now move our hands (and minds) and watch new versions of reality appear magically (and fast)? Not yet, partly because most of us don't have the mental focus to 'train the brain' to continually maintain a new thought/idea/paradigm.

In the difficulty of continuing to target and integrate unknown ideas is the fact that we don't believe such things are possible. Not only do we not find in our individual or collective mind, but we also do not feel in our bones.

We have not yet started to cultivate a culture in which quantum ideas are an obvious thing for our way of thinking. When humanity shifts its thoughts to reality, when the constructs and perceptions that take place collectively rotate 90 degrees, our material world can and will be a completely different place.

www.ingramcontent.com/pod-product-compliance
Lightning Source LLC
Chambersburg PA
CBHW070237220526
45465CB00004B/1447